HORNYHEADS,

MADTOMS, AND

DARTERS

Hornyheads, Madtoms, and Darters

Narratives on Central Appalachian Fishes

Stuart A. Welsh

OHIO UNIVERSITY PRESS

ATHENS, OHIO

Ohio University Press, Athens, Ohio 45701
ohioswallow.com
© 2023 by Ohio University Press
All rights reserved

To obtain permission to quote, reprint, or otherwise reproduce or distribute
material from Ohio University Press publications, please contact our rights and
permissions department at (740) 593-1154 or (740) 593-4536 (fax).

Printed in the United States of America
Ohio University Press books are printed on acid-free paper ∞ ™

31 30 29 28 27 26 25 24   5 4 3 2

Paperback ISBN: 978-0-8214-2610-4
Electronic ISBN: 978-0-8214-2611-1

Library of Congress Cataloging-in-Publication Data available upon request.
Names: Welsh, Stuart A., author.
Title: Hornyheads, madtoms, and darters : narratives on Central Appalachian
   fishes / Stuart A. Welsh.
Other titles: Narratives on Central Appalachian fishes
Description: Athens : Ohio University Press, 2023. | Includes bibliographical
   references and index.
Identifiers: LCCN 2023017328 (print) | LCCN 2023017329 (ebook) | ISBN
   9780821426104 (paperback) | ISBN 9780821426111 (adobe pdf)
Subjects: LCSH: Fishes—Appalachian Region. | Natural history—Appalachian
   Region. | Ecology—Appalachian Region—United States.
Classification: LCC QL628.A5 W45 2023 (print) | LCC QL628.A5 (ebook) | DDC
   597.0974—dc23/eng/20230803
LC record available at https://lccn.loc.gov/2023017328
LC ebook record available at https://lccn.loc.gov/2023017329

publication supported by a grant from

The Community Foundation for Greater New Haven

as part of the Urban Haven Project

To my parents, John and Janice,
whose unconditional love and encouragement
have provided endless opportunities, including
a path to pursue nature
and science

To my wonderful wife, Beth,
my daughters, Emily, Amelia, and Adrea,
and son, Evan, for their unconditional love
and unceasing support, and particularly
for their patience with
my overt obsession with fishes

CONTENTS

PREFACE

The central Appalachians, a nature-lover's paradise, happen to be a place I am quite fond of, and a place I call home. Perhaps the people are part of the appeal, but from a nature perspective, I have always pondered the diversity of life associated with an extensive network of waterways. The waters of small mountain streams converge, ultimately contributing to large lowland rivers. You have only to look beneath the surface of these waters to see an amazing dazzle of fish diversity on display. Some folk have lived long lives near these diverse rivers and streams, without looking and without seeing. Others look, but never see the beauty beneath the surface. This book is for those people willing to peer beneath the water's surface.

The contents of this book are a collection of stories on nature, naturalists, and the natural history of fishes. The book's focus is on many of the fascinating things that fishes do in their natural habitats. An ecological ensemble is played out from a species and population perspective, although community and ecosystem concepts are also present. These stories link central Appalachian fishes with concepts of competition and predation, species conservation, parasitic infections, climate change, public attitudes, reproduction and foraging ecology, unique morphology, habitat use, and native versus nonnative status of species. Nearly half of the families of central Appalachian fishes are represented, including the lampreys, gars, freshwater eels, pikes, minnows, suckers, catfishes, trouts, trout-perches, sculpins, sunfishes, and perches. The chapters in this book are arranged by phylogenetic order of fish families. Stories on more primitive fishes are presented first, beginning with the family Petromyzontidae (lampreys) and ending with the family Percidae (perches).

My stories often highlight old-school naturalists, individuals who contributed to our knowledge of nature during previous centuries. I emphasize references and excerpts from the writings of these naturalists, who wrote when science writing was less concise. These early naturalists provided a strong baseline of information, a knowledge base that underlies and supports much of our current science. Some would state this as "standing on the shoulders of giants," a concept that predates the life and times of Sir Isaac Newton. The old-school naturalists are all but gone, their writings are nearly forgotten, but I take great pleasure in bringing them back.

What defines the central Appalachians? You may be drawn to map the borders, but I was initially hesitant to draw boundaries. Simply put, the central Appalachians are between the northern and southern Appalachians. This beautiful region includes plateaus, mountains, and a Ridge and Valley region with a Blue Ridge border. I consider the northern edge of the central Appalachians to be near the southern extent of Pleistocene glaciation. The central Appalachians include eastern parts of Kentucky and Ohio, West Virginia, western Virginia, parts of southern and central Pennsylvania, western Maryland, and a small part of southern New York. During the publication process, reviewers asked for a map with boundaries. I have chosen to set these boundaries based on data from an old-school naturalist, Nevin Melancthon Fenneman (1865–1945). Dr. Fenneman, a native of Lima, Ohio, was a renowned geologist and geographer. In 1938, he published *Physiography of the Eastern United States,* for which he was awarded a gold medal by the Geographical Society of Chicago. My bounded central Appalachian map, in part, reflects his work on the eastern physiographic regions. On the map below, Fenneman's physiographic regions are generalized as the Appalachian Plateau, the Ridge and Valley, and the Blue Ridge. An artificial boundary of the southern border, marked on my map with a dashed line, reflects my uncertainty as to where the central Appalachians transition to the southern Appalachians.

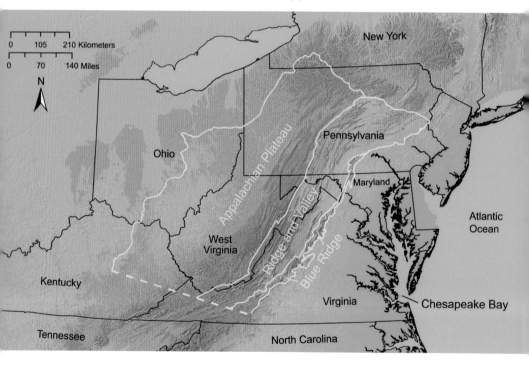

Some of my favorite science writers explain nature's complexities in simple, straightforward terms. I have tried to mimic this style of writing, but I am guilty of using my fair share of technical terms. I define science words upon first usage, and have included a glossary. Scientific names of fishes are emphasized in the book's chapters and listed in appendix 1. The scientific name of a species is a binomial composed of a genus and a specific epithet. Consider the scientific name of the Brook Trout—*Salvelinus fontinalis*. The genus is *Salvelinus* and the specific epithet is *fontinalis*. Also, please note that I follow the science convention of using the plural form "fishes" when referring to two or more species, and "fish" when referring to two or more individuals of the same species.

Many terms used throughout this book refer to a fish's morphology, such as a fish's body shape, its external and internal body parts, and the relationship between body parts. Words that refer to a location on a fish's body are dorsal, lateral, and ventral, or the general areas of the back, side, and underside, respectively. An anterior location on a fish's body is in the direction of the head, whereas posterior is toward the tail. Other fins include caudal, anal, pelvic, and pectoral fins. The caudal fin is the tail fin. Sometimes scientists splice words together. For example, the anal fin is located in a posterior and ventral position, or posteroventral position. The pelvic fin is in a ventral location. In the illustration below, the pectoral fin is positioned in an anteroventrolateral location. An adipose fin, located in a posterodorsal position, is present on some fishes of central Appalachia, such as catfish, trout, and Trout-perch.

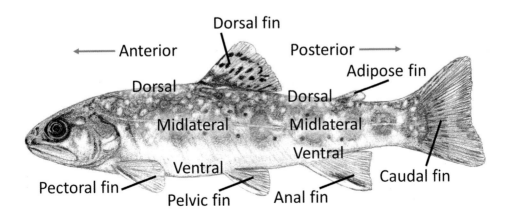

This book's conception, centered on the Appalachians, was influenced by nature and those who love nature, and particularly by the unbounded beauty of fishes. A naturalist's view of nature is nurtured by outdoor adventures, but books have a place. During my younger years I was inspired by several books, including one paperback that grabbed my attention: *Hen's Teeth and Horse's Toes,* by Stephen J. Gould. For me, this book bridged a gap, boosting a love of nature into bigger science thoughts. I hope that my stories on central Appalachian fishes will strike a chord with readers, boosting an interest in both nature and science. Most people are not scientists, but most people have an interest in nature, and thus in some sense are naturalists. Whether you are enamored with or only slightly interested in nature, I hope that this book will speak to your inner naturalist.

ACKNOWLEDGMENTS

I am indebted to many people who have helped me with the writing of this book. The insightful comments of George Constantz and Dustin Smith improved all chapters. Additionally, West Liberty University graduate students read and provided comments on chapters: Patrick Allison, Riley Aulik, Destinee Davis, Emmy Delekta, Garrett Hoover, Lane LeMasters, Greg Myers, Sydney Ozersky, Tess Prochaska, Nicole Sadecky, and Eric Tidmore. The Logperch chapter was improved by comments from Brendan Ebner. The background of the central Appalachians map in the preface is adapted from a color shaded relief map of the US Geological Survey. The figures preceding the opening page of chapters 5, 6, and 7 were adapted and redrawn from Jenkins and Burkhead (1994), courtesy of the American Fisheries Society. The opening illustrations of the Checkered Sculpin and Candy Darter chapters are based on photographs by Ryan Hagerty, US Fish and Wildlife Service. The opening illustration of the Logperch chapter is based on a video by Pat Rakes, Crystal Ruble, and J. R. Shute, Conservation Fisheries. Any use of trade, firm, or product names is for descriptive purposes only and does not imply endorsement by the US government.

1

Lamprey Enlightened

The central Appalachians support an amazing diversity of animal life, including secretive species that are seldom seen. In nature, stealthy behavior, often shaped by a species' evolutionary history, plays a part in the common ecological strategy for survival. Simply put, secretive animals try to avoid predators. Often, seldom-seen species are hiding during daylight hours and most active at night (nocturnal). Naturalists generally know where, when, or how to search for these animals. Perhaps Charles Conrad Abbott (1896) said it best in *Notes of the Night* with this nocturnal inquiry: "It is the purpose of Nature that man shall labor while it is day, and then rest; but the contrary is true of many millions of living creatures. These nocturnal toilers are not curious, in a human sense, as to man's doings, though influenced by them; but some of us are curious as to their ways, and so it is not strange that the question is often asked: What of the night?"[1]

Abbott was an avid naturalist, publishing books and scientific articles on many aspects of nature. He also served as a surgeon in the Union army, received a medical degree from the University of Pennsylvania in 1865, and was the first curator of the University of Pennsylvania's Department of American Archaeology.[2] One particularly important accomplishment was his 1860 species description of the Least Brook Lamprey,[3] a secretive, seldom-seen nocturnal toiler of the central Appalachians (figure 1.1).

FIGURE 1.1. Color sketch of an adult female Least Brook Lamprey, approximately 4.5 inches (11.4 cm) in length.

Relatively small-bodied, the eellike Least Brook Lamprey grows to about five inches (12.7 cm) in length, meal-sized for many piscivorous predators. From the life perspective of the Least Brook Lamprey, this species is low on the food chain, and presumably would be expected to have behaviors and adaptations that reduce its predation risk. In *The Rambles of an Idler*, Abbott (1906) wrote that "each creature knows its enemies and is ever on the alert to escape. To kill and not be killed is the burden of the single song sung by all creation."[4] Part of my curiosity about the Least Brook Lamprey revolves around a single question—how does predator avoidance influence the life history of this unique fish?

The Least Brook Lamprey, often referred to as a fishlike vertebrate, is fascinating, in part, because it differs drastically from most fishes. A long-listed explanation follows. First, it is jawless, with an elongate eellike body. Second, it does not have paired fins (pectoral or pelvic), but does have two dorsal fins, the posterior dorsal fin continuous with the caudal and anal fin. Third, seven gill openings are serially aligned on each side of the head, and a single nostril is found on top in front of the eyes. Fourth, the Least Brook Lamprey has a larval stage with poorly developed eyes, called an ammocoete (pronounced ăm' ō-sēt, meaning "embedded in sand").[5] The ammocoete has a hoodlike head with mouth parts adapted for filter feeding (figure 1.2). In general, lamprey larvae live in burrows created in soft stream bottoms,[6] and may live to about five years of age.[7] Fifth, ammocoetes metamorphose into adults, including a change of its hoodlike head covering to an oral disc (figure 1.2). Sixth, adults of Least Brook Lampreys are free-swimming, short-lived

FIGURE 1.2. (*left*) Ventral view of the hood-like head shape of an ammocoete (lamprey larva), showing specialized mouth structures associated with filter feeding, and (*right*) the oral disc of the mouth of an adult Least Brook Lamprey. Redrawn from Jenkins and Burkhead (1994), courtesy of the American Fisheries Society.

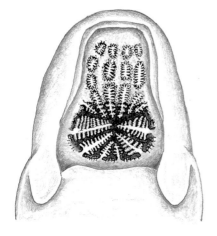 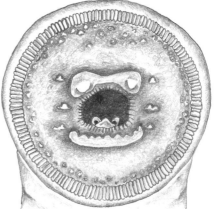

(< 1 year), and do not eat. Finally, adults congregate during the spawning period, often at the head of riffles, and die after spawning.[8]

Abbott described the Least Brook Lamprey based on a specimen from the Ohio River of the central Appalachians. During Abbott's time, this species likely had a wider geographic distribution, ranging from Ohio River to lower Mississippi River drainages, as well as several basins of the mid-Atlantic coast. This species is locally abundant in the Ohio River drainage of the central Appalachians and, as suggested by the name "brook," is most commonly found in creeks. Many stream populations have been reduced or extirpated because of habitat loss, likely owing to unnatural levels of stream sedimentation.[9] This species is unknown to most because of its burrowing behavior and nocturnal nature.

Other species of lampreys in the region also go unseen. Some lampreys, like Least Brook Lamprey, American Brook Lamprey, Northern Brook Lamprey, and Mountain Brook Lamprey, are nonparasitic. Others, like the Silver Lamprey and Ohio Lamprey, are parasitic species. Adults of the parasitic species use their oral discs to attach to other fishes to get a blood meal. Interestingly, the nonparasitic Northern Brook Lamprey and the parasitic Silver Lamprey are thought to be different morphs of the same species. A similar relationship is likely for the nonparasitic Mountain Brook Lamprey and the parasitic Ohio Lamprey.[10] A parasitic morph has not been found for the nonparasitic Least Brook Lamprey.

Some people consider lampreys to be lowly, with little value. In fact, as I noted previously, lampreys are an important prey species, providing food for aquatic and terrestrial predators. No one knows the complete list of predators that eat lampreys. In 1899, Professor H. A. Surface speculated that the posse of predators could include many mammals, such as "raccoons, muskrats, rats, minks, weasels, foxes and perhaps skunks and house cats."[11] Surface further suggested several bird predators, including hawks, owls, herons, bitterns, and shore birds, as well as reptiles and amphibians like water snakes, frogs, and mudpuppies. A more recent review of the literature supports adding insect larvae, hellbenders, turtles, ravens, mergansers, and domestic dogs.[12] I suspect that crayfishes and coyotes also eat lampreys. All piscivorous fishes co-inhabiting creeks with lampreys are likely lamprey consumers. The long list of predators reemphasizes the importance of lampreys as a link within and between aquatic and terrestrial food webs.

Least Brook Lampreys likely play an important role in stream nutrient cycling. Through filter feeding, a larval lamprey consumes very small food items, such as algae, protozoa, and other particles of suspended organic matter, thereby converting small energy sources into its own, larger body mass. Many aquatic animals are not able to use small particulate organic matter, but can consume larger prey, such as a lamprey. Further, adults undergo a short migration upstream, often to the nearest riffle, where they spawn and die, thus moving nutrients upstream within the watershed and slowing the downstream speed of nutrient spiraling.[13]

One of my colleagues, the late Philip Andrew Cochran, a naturalist and lamprey expert, lured me into the lives of lampreys through our many lengthy discussions. Predation was a shared interest. I recall several conversations on the brook lamprey's vulnerability to predation, including consideration of life stages, like adult, egg, and ammocoete.

In 1897, nest-constructing and spawning by brook lampreys were illustrated by Bashford Dean (figure 1.3),[14] and more recently generalized by Philip A. Cochran as follows:

> Lampreys most often spawn in streams during the spring. A typical lamprey nest is a depression in gravel substrate, usually at the upstream end of a riffle (i.e., just upstream from where the water surface is broken). Males typically initiate nest construction, but females

FIGURE 1.3. Brook lampreys on a spawning nest. Adapted from Dean and Sumner (1897).

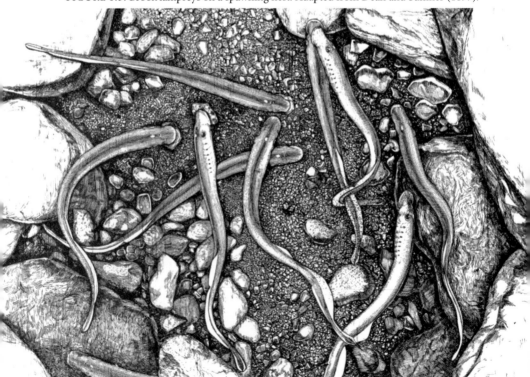

usually participate as well. Lampreys move stones or pebbles by attaching to them with their oral discs, lifting them from the bottom with thrashing motions, and letting the current help drag them downstream. They also attach to stones and sweep sediment by rapid lateral movement of their tails. Most species spawn in groups.

The typical mating act involves a female attached to a stone near the upstream end of the nest. A male attaches dorsally to her head and loops his tail around her body so that their genital openings are closely aligned. The female quivers rapidly as gametes are released. Mating is brief, lasting only a few seconds. Both sexes mate repeatedly, not necessarily with the same partners.[15]

A group of Least Brook Lampreys has a lot going on during the spawn, but concerns about predators are low on the list. First, spawning takes place during spring, a time when cold water temperatures suppress the metabolism and appetite of many larger fish predators. Second, most of the larger piscivorous fishes frequent deeper pool habitat of creeks, and generally do not forage in the moderate to swift-velocity shallow areas near riffle heads, the stream area where Least Brook Lampreys spawn. But birds, such as herons and kingfishers, are likely predators of spawning lampreys. Also, adult lampreys die soon after becoming spent from their spawning efforts, and predators likely gorge themselves during the downstream drift of dead or nearly dead post-spawn lampreys.

During spawning, predation of lamprey eggs is expected. In 1953, Seversmith published one of the few life history studies of the Least Brook Lamprey, positing some protection of eggs: "Upon fertilization the jelly about each egg expands. The egg falls to the bottom where it is soon covered by fine particles of gravel which adhere strongly to the extremely viscid outer jelly. The flurry of gravel caused by the spawning motions of the parents, as well as the movement of gravel caused by the changing flux of the stream, may cover the eggs and protect them."[16]

Nonetheless, egg predation is probably common. Small fishes such as minnows, sculpins, and darters are likely the largest, or at least the most common, fish predators of Least Brook Lamprey eggs.[17] As a countermeasure, females lay a lot of eggs, with some large individuals exceeding 5,000 eggs.[18] Given this high fecundity, I can't lament a few lost eggs, nor can the spawning female, as normal levels of egg predation have little impact on overall reproductive success.

Predation risk rises from when the eggs hatch to when small larvae find a place to hide.[19] Seversmith found that once the eggs hatch, the larvae "are either then caught in the changing water currents or swim to fine silt beds in very quiet (but not stagnant) water, and which are dark from the presence of thoroughly decomposed leaves and other organic matter. In this locale the developing ammocoete appears to stay throughout its first year."[20] If a larva eludes predators during its relocation to a leaf litter refuge, then its odds of survival look good. But this is a long journey for the little larvae, and many fall prey during this precocious period.

As the larvae mature, a habitat shift happens—individuals become susceptible to predation during a move from decomposed leaf litter to sand. But, by burrowing in sand, larval lampreys gain protection from many predators. Lampreys use mucus to solidify the walls of their burrow. These larval lampreys have a defined burrow, where they safely reside while filtering food from incoming water. Most piscivorous fishes, with the exception of the American Eel, would have difficulty extracting a larval lamprey from its burrow.

One of my graduate students researched habitat use of the larval Least Brook Lamprey.[21] Initially, he conducted an aquarium study where small and large lampreys were given a choice to burrow in six substrate types.[22] The small lampreys were less than two inches (50 mm) in length, whereas large individuals ranged from four to six inches (100–150 mm). Small ammocoetes preferred to burrow in fine sand of about 0.005–0.020 inches (0.13–0.50 mm) in diameter, a substrate selected by 55% of individuals (164 of 300; figure 1.4). Also, 31% (92 individuals) selected an organic substrate composed of about 70% decomposing leaves. Similar results were found for large ammocoetes, where the number and percentage for fine sand (149 individuals, 50%) exceeded the secondary selection of organic substrate (76, 25%). Some habitat categories were only occasionally used or rarely used, such as small gravel with diameter of about 0.09 to 0.19 inches (2.4–4.8 mm), coarse sand

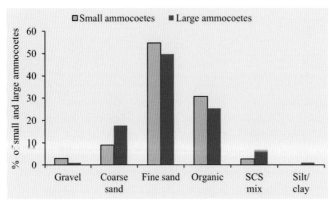

FIGURE 1.4. Small and large ammocoetes of Least Brook Lampreys, represented as percentages, that chose to burrow within six equally available bottom habitats (reanalysis of data from Smith 2009).

ranging from about 0.02 to 0.06 inches (0.5–1.4 mm), a mixture of silt and clay with small particles of less than 0.0025 inches (0.063 mm), and an even mixture of silt, clay, and fine sand (SCS mix; figure 1.4).

Habitat changes pose a problem for burrowed lampreys. During dry periods, when creek water levels drop, burrows can become dewatered, presenting opportunities for predators. If lamprey larvae can dodge predation during dry periods, then they may be able to survive dewatering, as individuals can obtain oxygen through their skin, as long as they remain moist. Alternatively, wet periods with high-velocity flood flows may scour lamprey beds, causing increased predation risk for relocating individuals.

Similar to larvae, burrowed adults remain relatively safe from predators. Larval lampreys transform into adults while inside their burrows, an important life history characteristic that reduces predation risk. Presumably, the longer the time period that transformed adults remain in burrows, then the less likely they are to fall prey to predation. Burrow emergence is inevitable as adults must eventually migrate upstream to spawn. Undoubtedly, predation risk rises during the migration, although the upstream journey often involves a brief period over a short distance.

In some cases, ammocoetes leave their burrows in response to stream habitat alteration. In an all too common example of benthic habitat alteration in central Appalachian streams, large amounts of silt and clay particles enter the stream after being dislodged by land disturbances. The small particles of clay and silt that enter the stream with rain runoff during storm events can blanket the bottom of streams. Sedimentation may stimulate lampreys to emerge from their burrows, search out better habitat, and ultimately risk higher predation during the process.

In another study published in 2012, D. M. Smith and colleagues used an experimental approach to examine the relationship between predation risk and availability of preferred habitat.[23] In that study, we considered that if preferred habitat is present, then ammocoetes may remain burrowed and be less susceptible to predators. In contrast, if the preferred habitat is not available or altered, then lampreys may spend (1) more time searching for better habitat, or (2) more time attempting to burrow in least-preferred habitat. If lampreys are swimming around searching for preferred habitat or spending a lot of time attempting to burrow into substrates, then predation risk will likely increase.

The predation experiment involved three substrate types: coarse sand, fine sand, and a 1:1 silt/clay mix. Smith placed 20 large ammocoetes into an aquarium with coarse sand, 20 individuals into an aquarium with fine sand, and 20 into an aquarium with silt/clay substrate. The lampreys were given 24 hours to acclimate and burrow into the aquarium substrate before two large Yellow Bullhead Catfish (11.5–12.5 inches, or 29.2–31.8 cm) were placed into each tank. After six days, the substrate trays were removed from the three aquaria and the number of remaining lampreys was counted. The entire process was repeated ten times.

On average, predation mortality (20%) was lowest when ammocoetes were able to burrow into their preferred habitat of fine sand (figure 1.5). In contrast, predation mortality (96%) was highest when larval lampreys were allowed to burrow into silt/clay habitat. We also observed that larvae took a longer time to burrow into silt/clay than fine sand, and we noted that individuals often emerged from silt/clay burrows at night but rarely left a fine-sand burrow. Based on our experimental results, we speculate that wild lampreys are likely more vulnerable to predators when preferred habitat is not available—a condition that can result from stream degradation owing to clay/silt sedimentation of the stream bottom.

In my mind, the Least Brook Lamprey is a beautiful, unequivocal evolutionary marvel. "No natural object can be ugly, repulsive, uninstructive, or

FIGURE 1.5. Least Brook Lampreys eaten by Yellow Bullheads, estimated as percentages, based on ten experimental trials and three habitat types. On average, predation mortality was considerably less when ammocoetes were in aquaria with fine sand substrate (20% mortality) relative to that for aquaria with substrates of coarse sand (42%) and silt/clay (96%; data from Smith 2009).

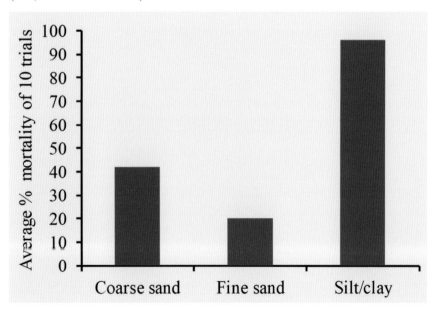

unentertaining," Charles Conrad Abbott revealed in *Nature's Realm* (1900), "if we see it as it is, and have knowledge of its place and purpose."[24] The problem with this, at least with the Least Brook Lamprey, is that relatively few people know the place or purpose of this reclusive species. Hopefully, this story helps by piquing the interest of those with new knowledge of this small nonparasitic lamprey.

Perhaps as naturalists, if we become nocturnal toilers, then we can learn more about nocturnal species. In *Outings at Odd times*,[25] Abbott (1890) wrote that "much is overlooked if we familiarize ourselves only with the events of the day, and ignore, as young naturalists are all too apt to do, the night-side of nature." Yet, even at night, the filter-feeding Least Brook Lamprey larvae are in burrows, concealed from both predators and the prying eyes of knowledge-seeking naturalists. Aquarium-based studies may be the best bet for improving our knowledge about this species. While pondering these future studies, I predict that Charles Conrad Abbott would marvel at the population persistence and predator avoidance of the Least Brook Lamprey, a remarkable species that so often goes unnoticed and unappreciated in our central Appalachian streams.

Ill-Regarded

Attitudes about nature are often nurtured during our younger years, including perceptions about predators. From some of my earliest observations, I realized that predator-prey relationships were an innate part of nature. Even so, I still developed a prejudice against predators. For some reason, predatory animals, particularly those with large teeth, are easily vilified, viewed as voracious villains or varmints. Perhaps we are indoctrinated early in life through children's books and cartoons. In my time, the culprits included such classics as Little Red Riding Hood, Wile E. Coyote and the Road Runner, and Ralph Wolf and Sam Sheepdog. A particularly memorable influence was the voice of Orson Welles narrating the Rudyard Kipling classic "Rikki-Tikki-Tavi"—the mongoose was portrayed as the good guy, and a hissing, slithering snake as the evil villain. Regardless of the reasons, like many children of central Appalachia, I developed a negative attitude toward predators, especially snakes.

During my childhood I often stayed with my grandparents during summer, spending as much time as possible catching critters in the backyard creek. I clearly remember my grandmother saying, "Let me know if you see a big snake in the creek." She was apparently not interested in little snakes. Several times each summer it happened. "Hey, Grandma," I would shout. "There's a big snake in the creek." Within seconds, she would appear from the backroom carrying the beautifully crafted Remington Targetmaster, Model 510. With its open sights and a .22 short, I never saw her miss a head shot on a Common Water Snake. My grandmother was a wonderful person; she just had a dislike for snakes, a central Appalachian commonality.

My grandma never read Aldo Leopold's *A Sand County Almanac*—but I did. It influenced my thinking about predators. Leopold, a renowned naturalist and the father of wildlife management, wrote the essay "Thinking Like a Mountain," in which he described the shooting of a wolf:

> We reached the old wolf in time to watch a fierce green fire dying in her eyes. I realized then, and have known ever since, that there was something new to me in those eyes—something known only to her and to the mountain. I was young then, and full of trigger-itch; I thought that because fewer wolves meant more deer, that no wolves would mean hunter's paradise. But after seeing the green fire die, I sensed that neither the wolf nor the mountain agreed with such a view.[1]

Some predatory fishes, particularly those with large teeth, are often treated like snakes and wolves, disliked and disdained by many people. Consider the Longnose Gar (figure 2.1), a long ill-regarded and often disrespected predatory fish of central Appalachian rivers. But why such a long-lasting ingratitude?

I find the Longnose Gar to be beautiful, with its elongate, streamlined body, attractively spotted fins, a unique "needlenose" snout, and an armoring of unusually hard body scales. The Longnose Gar is a predator. Its dorsal and anal fins are located posterior on the body near the caudal fin, a character similar to some other predatory fishes, such as Northern Pike and Musky (both of the genus *Esox*).

The Longnose Gar is sometimes colloquially called a Gar Pike. Carolus Linnaeus provided the species description in 1758, naming it *Esox osseus*.[2] In 1803, the French naturalist Bernard Germain de Lacépède listed the genus name as *Lepisosteus* and emphasized the unusual scales as "body and tail coated with very large scales, placed one above the other, very thick, very hard, and of bony nature."[3] *Lepisosteus* is from the Greek words *Lepis* (scale) and *osteon* (bone). The epithet *osseus* is from the Latin word *os* (bone). The Longnose Gar is widely distributed, residing in the central and eastern United States throughout much of the Mississippi River watershed basin, as well as parts of the Great Lakes, Atlantic Coast, and Gulf Coast drainage systems. In the central Appalachians, this fish is common within the Ohio River and its larger tributaries. The fish is also quite common in the lower reaches of rivers that drain the eastern slope of the central Appalachians.

Perhaps the Longnose Gar is loathed because of its looks, often compared to those of a snake. The Longnose Gar is most certainly a fish, but

FIGURE 2.1. A color sketch of a Longnose Gar (scientific name *Lepisosteus osseus*).

its elongate narrow body and atypical hard scales appear to some as more reptilian than fishlike. In his monograph on fossil fishes (1833–44), the Swiss American naturalist Louis Agassiz stated that "their tegumens [body armor] often have an appearance so similar to those of Crocodiles, that it is not always easy to distinguish them."[4] Agassiz was one of the first scientists to study these scales, which he called "ganoid" scales. "The most common structure of Ganoid scales is that of a bone crest covered with a layer of enamel," noted Agassiz, further stating that "the enamel layer is especially striking" because "it is perfectly separated from the bone substance and formed of a material hard, brittle, transparent and without apparent structure, similar to a layer of glass."[5] The ganoid scales of the Longnose Gar are interesting and complex, as detailed by a subsequent study published in 1849 by the English naturalist William Crawford Williamson (figure 2.2).[6]

Along with a general disdain for them as toothy predators, gars are often perceived as a threat to "desirable" species such as food fishes and sport fishes. "Because they feed upon or compete for food with more desirable fishes," noted Pflieger, "gars are often considered to be a worthless nuisance."[7]

FIGURE 2.2. Vertical section of a ganoid scale of a Longnose Gar (A). The scale surface is covered with a thin layer of enamel, called ganoine. Layers of the bony scale are intercepted by canals (tubes), some of which are narrow, entering from the top and bottom of the scale, or passing longitudinally, whereas others are wider and passing completely through the scale. Part of the anterior scale is embedded and overlapped by the posterior portion of the adjacent scale. Each scale interlocks with surrounding scales, so that the exposed portion of each scale has a rhomboid shape (B).

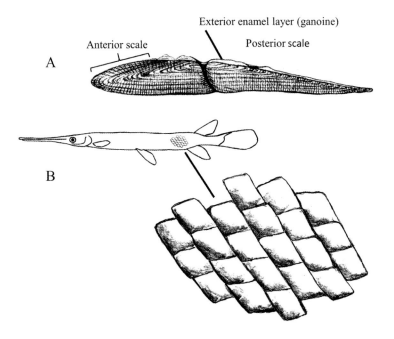

FIGURE 2.3. A 1956 photograph of a catch of Longnose Gar from Fish Creek, a tributary of the Ohio River. Reprinted from Van Meter (1956), courtesy of West Virginia Division of Natural Resources.

Gars have long been slaughtered in great numbers. For example, Jenkins and Burkhead observed people "gathered at bridges in southwestern Virginia to shoot gars with guns during spawning runs."[8] In 1956, Van Meter commented on a central Appalachian gar-gigging party on a tributary of the Ohio River (figure 2.3): "Saturday, January 14, three men went down to Meighen Bridge at Fish Creek and joined with the landowner in a gar-gigging party. Mostly within a couple of hours, they had landed 450 of the big fish."[9]

Negative attitudes about Longnose Gar were aided by the early accounts by some ichthyologists. For example, in the 1920 book *The Fishes of Illinois,* Stephen Alfred Forbes and Robert Earl Richardson stated that gars were "as useless and destructive in our productive waters as wolves and foxes formerly were in our pastures and poultry yards."[10] The authors further stated:

> This voracious, active, and well-protected fish is a notable winner in the long struggle for existence which its species has maintained, but it is a wholly worthless and destructive nuisance in its relations to mankind. It is the enemy of practically all the other fishes in our waters, and so far as it eats anything but fishes, it subtracts from the food supply of the more valuable kinds. It has, in fact, all the vices and none of the virtues of a predaceous fish. On the other hand, it is preyed upon by nothing that swims, and is so well adapted to the varied features and vicissitudes of its habitat that it is proof against any but the most extraordinary occurrences.[11]

Interestingly, Forbes and Richardson, despite a general dislike for adult gars, fell prey to Konrad Lorenz's "baby schema" effect, stating: "They are extremely interesting, and even beautiful, little animals, each marked with a

FIGURE 2.4. A juvenile Longnose Gar about three inches (76 mm) in length. Note the interesting elongate upper lobe of the caudal fin, a peculiar character not present in adults.

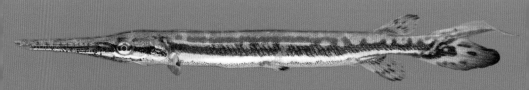

broad black lateral band; and they are especially noticeable for the evanescent lance-shaped upper lobe to the caudal fin"[12] (figure 2.4). Lorenz's "baby schema" was based on the attractiveness of an infant's physical features, including a big head and large eyes.

Early development and larval stages of the Longnose Gar were studied as early as 1878 by Alexander Agassiz, son of Louis Agassiz.[13] His work demonstrated another extremely interesting characteristic of larval Longnose Gars—reverse countershading (figure 2.5). Most fishes are countershaded as dark dorsally and light ventrally. This pigment pattern provides concealment from predators, because the dark dorsum is lightened by sunlight and the light ventrum is darkened by shadows (see chapter 18, "Like a Rolling Stone"). The ichthyologists Herb Boschung and Richard Mayden suggested that reverse countershading may give motionless young Longnose Gars the appearance of being dead or "belly up," perhaps putting the larvae in a position to be close to their prey species.[14]

In 1954, under the heading "Gar Control as an Instrument in Fish Management," Holloway wrote: "Gar control has been considered a method of improving waters for angling by some state conservation agencies and by sportsmen, (Lagler et al. 1942, Gowanloch, 1939 and others). This popular conception has been given some justification by this and other studies showing gars to be highly piscivorous, eating fishes of greater value to man than are the gars; and competing for food with predaceous game-fishes."[15] But later in his paper he questioned the need for gar control, mulling over the reasoning for culling gar based solely on their predatory nature. Holloway hedged:

> It remains to be demonstrated clearly whether the reduction of gars will result in an increase or decrease in the harvest of desirable species. Neither has it been shown that gars act as regulators to prevent overpopulated or unbalanced populations, but this has been assumed by most workers due to the piscivorous habits of gar. Lagler et al. (1942) point out among other facts that ". . . gar populations are greatest (and perhaps most useful) where the largest populations of buffer, forage and coarse species of fish occur;" and that "it is perhaps logical to

FIGURE 2.5. A "cute," reverse-countershaded larval Longnose Gar about one inch (2.5 cm) in length. Adapted from Agassiz (1878).

expect that, if not controlled in some way their numbers will increase as man goes on upsetting the balance to favor them by cropping those predaceous food fishes that presumably are their natural enemies."[16]

In a 1914 paper titled "Fishes Swallowed by Gar Pike," Hussakof wrote: "As is well known, gar pikes are highly predacious fish. They devour vast numbers of food and game fishes; and in localities where they are abundant they are treated as pests and destroyed by the thousand."[17] Studies on the diet of the Longnose Gar, however, have generally not supported that Longnose Gar "devour vast numbers of food and game fishes." Interestingly, many studies have reported a low percentage of game fishes in the diets of Longnose Gars. For example, in 1940, Karl F. Lagler and Frances V. Hubbs found a frequency of occurrence of 25% game fishes (mostly species of bullhead catfishes, sunfishes, and Yellow Perch) and 72.1% other fishes (mostly species of minnows, madtoms, and darters).[18] Another study reported a diet composition of 7% game fish, 69% forage fish, and 24% fish remains.[19] The diet of the Longnose Gar, like that of most generalist predatory fishes, is determined by availability and abundance of prey species. If small-sized food and game fishes are extremely abundant, then a Longnose Gar will likely include these fishes in its diet.

Given that gars are armored, it seems only fitting that they were a study subject of Bashford Dean. A late nineteenth-century renaissance man, Dean simultaneously held positions of professor of vertebrate zoology at Columbia University, curator of reptiles and fishes at the American Museum of Natural History, and curator of arms and armor at the Metropolitan Museum of Art.[20] In 1895, Dean acknowledged that gars were not well liked: "Belief, however, is very current that Gars are exceedingly voracious, killing and eating the larger fish, and not hesitating to attack even a swimmer who has ventured in their neighborhood."[21] But he then provided an interesting account that dispelled these myths:

> The actual mode of feeding has been observed by the present writer. The Gar approaches its prey (young dace) cautiously, advancing without perceptible movements; when within three or four feet it pauses, as if accurately to direct its aim, then, without seeming effort, it shoots quickly forward, secures the fish, stops suddenly, and is again motionless. An occasional bending of the head adds not a little to the apparent dexterity of movement. The food in all cases examined consisted exclusively of small soft finned fishes, mainly dace, none of which were longer than three and a half inches; young perch and sunfish,

abundant in the locality, do not appear to be eaten. The number of small fishes each Gar had taken was especially large and fully justifies the idea of the fish's rapaciousness; from the stomach of a male (24 in.), caught while spawning, eleven cyprinoids were taken; of another (27 in.) the stomach contained the remains of thirteen fishes, while in addition three (of 2 in., 2 ½ in.) were taken from the pharynx. From all observations it would appear that the Gar is reasonably to be looked upon as only indirectly injurious to food fishes—*i.e.,* bass, perch, pickerel, pike, catfish—in reducing the general food supply. It is also worthy of note that there appeared throughout no evidence of the fish's having taken food that had been torn or cut; the function of the straight, close set teeth seemed rather to prevent the escape of the prey than to kill or cut it.[22]

Many people do not realize the important role that predators play in nature. As Leopold stated, "I thought that because fewer wolves meant more deer, that no wolves would mean hunter's paradise."[23] Similar errant reasoning is often used about predator-prey balance of fishes and angling success. But as reported by Robison and Buchanan, "Gars play an important role in preventing overpopulation of many other species and in maintaining a proper natural balance in many natural lakes and impoundments."[24] Similarly, Clay noted that "predators generally improve the populations of their prey species through their selective harvesting,"[25] and Pflieger stated that gars "may serve as a natural control in preventing overpopulation and unbalanced fish populations."[26]

In closing, consider the wise words of Alfred Cleveland Weed: "Man is so constituted that he considers the value of other living things solely on the basis of his own comfort or convenience. A fish is useful or valuable to him only as he can see some direct relation to his needs or pleasures. On this basis, much has been said against the gars and very little in their favor."[27] Fortunately, negative attitudes about predators appear to be improving. In a paper titled "Changes in Attitudes toward Animals in the United States from 1978 to 2014," George et al. reported that attitudes have changed about some historically stigmatized predatory species. Specifically, positive attitudes on wolves and coyotes increased by 42 and 47%.[28] Somehow I doubt that such an increase has occurred for the Longnose Gar. But I remain optimistic and in agreement with Scarnecchia, who wrote that "our waters (and we ourselves) may be much better off ecologically and aesthetically with a diverse fish community including gars."[29]

Eel Ways

It would be an understatement to label the American Eel as unusual—a fish mysterious to many, yet unknown to most. American Eels may look and move like a snake, yet they are very much a fish with fins and gills (figure 3.1). My first eel experience was with hook and line: a friend reeled in an eel while night fishing on the Potomac River near Paw Paw, West Virginia. The eel squirmed and slithered on the shore, sliming its path in the sand. As young boys, we were uncertain about eels. Maybe it was electric, or possibly its tail was poisonous, or it could strike with snakelike fangs. I cut the line close to the hook, and the eel, moving like a snake, returned to the river. Within 15 minutes, I reeled in an eel. "Mine is much bigger than yours," I boasted, with both a competitive boyhood tone and a slight fisherman's exaggeration. On a closer look, we noticed the cut line and hook from the previous catch—it was the same eel. My bragging rights were silenced. The eel was returned to the river, never to be seen again by us, but never to be forgotten. The eel imprinted a lifelong memory, one that started as simple curiosity and later matured into a research obsession.

My research focus often flows to rivers and their fascinating fishes. Rivers are the road networks of nature—the aquatic highways of the Appalachian Highlands. Rivers run from the central Appalachians to the ocean, a rough and rocky road with highland waterfalls and steep gradients descending to coastal lowlands. Most rivers draining the central Appalachians are

FIGURE 3.1. A yellow-phase American Eel from the Shenandoah River.

dammed, often multiple times along the way from highlands to the ocean. On the western slope of the central Appalachians, rivers run toward the lowlands of the Mississippi River Valley and seaward to the Gulf of Mexico. On the eastern slope, rivers descend to the Atlantic Ocean.

The connection between highland and ocean is critical for the American Eel. American Eels hatch in the ocean, but some swim long distances up rivers during adolescence. Adults travel back downstream to the ocean to spawn. Outside of the central Appalachians, migratory salmon experience a somewhat similar connection between rivers and ocean. Salmon migrations are opposite to those of the American Eel. Salmon mature in the ocean and spawn in rivers, and eels mature in estuaries or freshwater rivers and spawn in the ocean. Based on the spawning migration, salmon are labeled as anadromous, whereas eels are called catadromous. Since catching my first eel on the banks of the Potomac, I have often wondered about the long-distance river migrations of catadromous American Eels. How do eels swim up and down dammed rivers?

American Eels are not the only fishes that move up- and downstream in river systems—some small fishes, such as sculpins, move short distances, and larger fishes, like paddlefish and sturgeon, travel long distances. In fact, all fishes move upstream and downstream within waterways, but no fish within the central Appalachians travels as far as the American Eel. Waterfalls are often barriers for upstream fish movements, although most waterfalls within central Appalachian watersheds are found in high-elevation headwaters, in streams too small for American Eels. Farther downstream, the steep slopes of many central Appalachian rivers create sections of fast-flowing water called cascades, riffles, or rapids, but most fishes, especially eels, are capable of moving upstream and downstream through these areas. The native fishes of the central Appalachians have lived with steep gradients for a long evolutionary time. Earthen or concrete dams, however, are relatively new to the central Appalachians and its fishes. Fishes do not deal well with dams, which often prevent upstream movements, and sometimes even downstream movements.

Dams are useful to us, but not to many fishes. They are widespread throughout watersheds of the central Appalachians. The National Inventory of Dams (NID) lists a total of 6,717 dams for the states of Kentucky, Maryland, Ohio, Pennsylvania, Virginia, and West Virginia. Most of these dams are less than 50 feet (15 meters) high; 44% are less than 25 feet (7.6 meters), 41% are from 25 to 50 feet (7.6 to 15.2 meters), 10% are from 51 to 100 feet (15.5 to 30.5 meters), and 5% are greater than 100 feet (30.5 meters). About half of those dams were built during 1950 to 1980. The total number of dams is much higher, because the NID count does not include dams less than 6 feet (1.8 meters) high with less than 50 acre-feet (61,674 cubic meters) of water storage. Dams are built for various reasons, such as flood control, navigation, hydropower, water supply reservoirs, and recreational activities such as boating, fishing, and swimming. In the central Appalachians, most dams function as flood control, and most are used for some recreational activities. Navigational locks and dams occur on the Ohio River (Greenup, R. C. Byrd, Racine, Belleville, Willow Island, Hannibal, Pike Island, New Cumberland, Montgomery Island, Dashields, and Emsworth), the Allegheny River (lock and dam 2, C. W. Bill Young, and locks and dams 4, 5, 6, 7, 8, and 9), and the Monongahela River (Braddock, lock and dam 3, lock and dam 4, Maxwell, Gray's Landing, Point Marion, Morgantown, Hildebrand, and Opekiska). Some examples of hydropower dams occur on the Cheat River (Lake Lynn Dam), Potomac River (dams 4 and 5), and Shenandoah River (Millville, Warren, Luray, Shenandoah, and New Port).

Native Americans and European settlers constructed weir dams, a wall of stacked rocks crossing a river in a *V* shape, which funneled fishes into enclosures for use as food. Weir dams were common within Atlantic slope drainages from the eastern side of the central Appalachians to the Eastern Seaboard. From aerial photography of the Potomac River, 39 weir dams are documented from the lower end of Heater's Island upstream to Harpers Ferry.[1] These historic weir dams are now in an eroded condition. Most are considered to be of colonial construction based on the straight V-shaped wings, and others are attributed to Native American origin based on irregular wings. Possibly most were originally Native American origin and later reconstructed by colonists.

In later years, during the late 1700s and early 1800s, dams were built to divert water into canals, manmade ditches parallel to rivers for transport of goods. George Washington and Chief Justice John Marshall were major players in the push for these canals to connect the Eastern Seaboard with lands west of the Allegheny Mountains. The James River and Kanawha Canal followed the James River from Richmond, Virginia, upstream to Buchanan, Virginia, and include 160 miles (257.5 km) of canal, 37 miles (59.5 km) of river navigation, 90 locks, 23 feeder dams, and 12 aqueducts.[2] Another canal system, the Chesapeake and Ohio Canal, parallels the Potomac River for 185 miles (298 km) from Washington, DC, upstream to Cumberland, Maryland. This canal, in operation from 1831 to 1924, was associated with 182 culverts, 11 aqueducts, 74 locks, and 6 dams.[3]

On the western side of the central Appalachians, wicket dams were constructed for boat navigation in the Ohio River and its larger tributaries during the late 1800s and early 1900s.[4] Initially, a series of wooden wickets was hinged to the river bottom and raised during periods of low water flow. In later years, the movable dams were replaced by concrete lock dams, which currently allow barge transport of commercial goods.

Undoubtedly, dams change rivers. For one thing, rivers are free-flowing waterways—dams slow the flow, creating lakelike conditions upstream. Some biologists believe that these changes have contributed to recent declines in native fish diversity, where both the number of fish species and the number of individuals of some fish species have declined. For American Eels, dams are like roadblocks—sometimes eels get by them, and sometimes they do not.

Unlike eels, most fishes of the central Appalachians are born and die in central Appalachia. But American Eels become temporary residents of the central Appalachians after swimming upstream from either the Gulf of Mexico or the Chesapeake Bay. From the Gulf of Mexico, an eel reaches the central Appalachians by swimming up the Mississippi River, and then up the Ohio River and its larger tributaries, where thousands of smaller streams branch off in upstream directions. From the Chesapeake Bay, eels enter the central Appalachians by swimming up the James or Potomac River. The Susquehanna River was previously an important path for eels, but since 1928 the Conowingo Dam on the lower Susquehanna River has prevented eels from migrating upstream. Efforts are now underway to return eels to the Susquehanna. No one knows the exact number of eels within the central Appalachians, although the number is in flux, as some eels move upstream and others move downstream. The numbers of eels have also declined in recent years within the central Appalachians, but this population decline has been reported for the entire range of the American Eel—for many, it is a cause for concern.[5]

To understand the mysterious American Eel, one must realize that it is nothing like a usual fish. Amazingly, American Eels begin life as fertilized eggs in the Sargasso Sea, an area of the Atlantic Ocean that spans Bermuda to the Azores. Consider the many life changes of the American Eel from egg to adult (figure 3.2). The larva, called a leptocephalus, is shaped like a willow leaf. Larvae are not good swimmers, but drift with ocean currents from several months to a year, while growing to a length of about 2.0 to 2.5 inches (5.1–6.4 cm). The larva transforms into a transparent eellike body, when it is known as a glass eel. Glass eels, ranging in size from 2 to 4 inches (5.1–10.2 cm), enter coastal estuaries by swimming or using tidal currents. Glass eels darken with body pigments and are then referred to as elvers, and later as

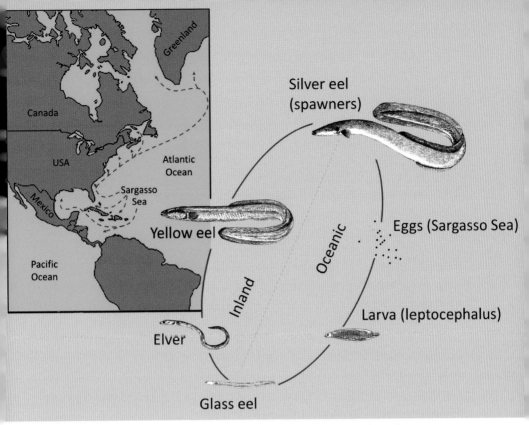

FIGURE 3.2. Life cycle of the American Eel and predicted pathways of larval transport.

yellow eels. Though size marking the change from an elver to a yellow eel is somewhat arbitrary, some biologists use a cutoff length of about 4 inches (100 mm) to differentiate the two. From the estuaries, many elvers begin swimming up rivers, a journey continued by yellow eels.

Sometimes one must expect the unexpected—especially with eels. Some yellow eels stay near estuaries, the coastal areas where salty ocean water mixes with the freshwater of rivers. Many of the eels that stay within estuaries or near river mouths are males. Some yellow eels ascend rivers, some not venturing too far upstream, whereas many eventually reach the river's headwaters. Most of the yellow eels within rivers are females, while those ascending to headwaters are always females. Within a watershed, the female yellow eels are often larger at farther distances from the ocean. Yellow eels encounter obstacles while moving upstream, but can easily move through the fast-flowing riffles of rivers. Dams are different. Some dams are dead ends for eels, whereas others temporarily detain upstream movement. With trial and error, eels often find a way over or around, though it is generally unknown how they do it. Eels are nocturnal—mostly moving during the darkness of night, making it difficult to observe their movements.

Some yellow eels ascend short distances up rivers, but others migrate long distances upstream to headwaters. In larger watersheds, like the Potomac River on the Atlantic slope or the Mississippi River on the Gulf slope, an eel's upstream journey may take several years to complete. Most long-distance movements upstream occur when river flows are rising from recent rains. In the central Appalachians, higher rainfall and river flows commonly occur during spring or with hurricane-induced storm events during fall. Yellow eels often do not migrate upstream during low flow periods of summer. They also stop moving upstream during late fall and winter when water temperatures drop below 50°F (10°C). Eels often overwinter near tributary mouths and spring seeps, or burrow within soft sediments of the stream bottom, where water temperatures may be a few degrees warmer than the surrounding environment.[6]

No one knows why some eels swim far upstream and others do not. If all eels converged to a common area, though, competition would be extreme for space and food. Many eels migrate to larger headwater streams, where some reach lengths of around 40 inches (one meter). Most small headwater streams are too small for large eels. In my experience with central Appalachian headwater streams, if you can step across the stream, then it is likely too small for large eels.

Within headwater streams, eels hide in many habitats during the day and actively search for food at night. I have found eels within rock crevices, under large rocks, under submerged root wads and logs, within leaves on the bottom of the stream, and burrowed into soft sediments. Eels eat crayfish, other invertebrates, and small fish. Most fishes cannot eat large prey that exceeds their mouth-gape size. Eels, however, by biting and pulling or biting and spinning, can tear off small pieces of large prey. After grasping a large prey item, an eel can pull backwards with substantial force. An eel can swim backward as well as it can swim forward, an ability known as palindromic swimming. Further, an eel can clamp its jaws on larger prey and pull off a piece of meat by spinning its body at up to 14 spins per second; this is called rotational feeding.[7]

Near the end of their lives, yellow eels go through a final transformation to a silver eel. They change color from a yellowish back and sides to a darker bluish-silver back and silver sides. Silver eels also develop larger eyes, thicker skin, and more muscle mass. Around this time, the eel starts its downstream journey to the Sargasso Sea.[8] Along the Atlantic Coast, American Eels are influenced by latitude—lifespan and length of adults decrease from north to south.[9] Also, females live longer and grow larger than males.[10] In Canadian rivers of Quebec and Nova Scotia of the northern range, the average age and length of adult silver females are 20.4 years and 31.5 inches (80 cm). Silver males in the northern range are smaller than females and average 12.9 years of age and 14.5 inches (36.8 cm). In Atlantic coastal rivers of the southern United States, adult silver females average 7.5 years old and 23.2 inches (58.9 cm) in length, whereas the smaller males average 4.2 years and 14.2 inches (36.1 cm) in length. The latitudinal difference in size and age at maturity may be explained by slower metabolic rates of eels in colder climates of the northern latitudes. Very little information is known about eels from the rivers and bays of our mid-Atlantic region,[11] the headwaters of which start in the central Appalachians.

Despite differences in size and age of silver eels, the Sargasso Sea is the desired end destination for all silver eels as they swim back to the sea to spawn (and die). The seaward migration of silver eels often begins in the latter part of the year, from September through December. Many do not survive the journey. Some succumb to predators, others are caught by commercial fishermen. Death by dam is another possibility. For rivers with dams, young yellow eels successful at getting upstream must again face the dams on the way downstream as adult silver eels. Most eels within headwaters must pass dams for a successful downstream journey. Eels can either follow the water over the dam spillway or follow the flow through the dam outflow. Hydropower dams pose an additional risk if silver eels follow the downstream flow toward the turbines. Hydropower turbines spin with water flow and are known to crush and chop up eels.

My research has focused on the upstream movements of yellow eels and the downstream movements of silver eels, primarily within the Potomac River watershed of the Chesapeake Bay. Once entering the Chesapeake Bay from the ocean, a young eel may stay within or near a river mouth, or swim up one of many rivers. My research has focused on those eels that swim up the Potomac River.

Young eels swimming up the Potomac pass along the southwestern border of Washington, DC, and within a few miles cross a dam at Little Falls that is 12 feet (3.66 m) high. Approximately eight miles (12.9 km) farther upstream, the eels cross the Fall Line, which includes a landmark known as the Great Falls of the Potomac—a steep cascade that separates the coastal plain and the Piedmont plateau. The eels encounter another dam at the top of Great Falls, one that is 10 to 15 feet (3.0–4.6 m) high and diverts water into the intake of the Washington Aqueduct—the public water supply of the nation's capital. About 8 miles (12.9 km) upstream of Great Falls, the eels cross a rubble dam at Violette's Lock. The eels travel another 38 miles (61.2 km) upriver before reaching the confluence of the Shenandoah River at Harpers Ferry. At this juncture, eels take either a left-fin turn up the Shenandoah River or continue up the Potomac. Those swimming up the Shenandoah have contributed most to my knowledge of eels.

The Shenandoah, like the lower Potomac, is dammed. From the Shenandoah's mouth, an eel travels a mere 4.5 miles (7.2 km) before reaching a dam near Millville, West Virginia (figure 3.3). Millville Dam was built for hydropower, and the dam includes the gated entrance of a canal on the west bank that leads downstream to a hydropower station. At a height of about 17 feet (5.2 meters), the 971-foot-wide (296-meter-wide) Millville Dam is only a temporary barrier to many American Eels.

During the summer of 2003, a fishway was installed on Millville Dam to aid the upstream passage for American Eels. Fishways, sometimes known as fish ladders, are ramp or steplike structures that allow fishes to pass over a dam. The fishway at Millville, or in this case "eelway," is a 36.5-foot-long (11.1-meter-long) covered metal ramp with a 50° slope on the western end of the dam. It seems unlikely that a yellow eel moving upstream and approaching the dam would immediately find the eel ladder. Instead, eels likely swim along the dam face with unsuccessful attempts to swim up the flow of the falling water. At the eelway, the flow of water over the top of the dam is stopped by flash boards, creating a pool of still water at the base of the ramp. Water from inside of the eelway and from an adjacent gravity-fed pipe provides an attractive flow. Finding the bottom of the eelway, an eel begins to swim up the ramp. Inside the eelway, small pipes are positioned as a "peg board," a climbing substrate well suited for the serpentine movements of eels. After climbing to the top of the eelway on the downstream side of the dam, eels swim through a large pipe to the upstream side. After reaching the upstream side of the dam, they continue down the pipe like an amusement park ride for about 150 feet (45.7 meters) of twists and turns until dropping into a collection tank. Initially, a net bag was used, where eels were lifted out of the collection tank, counted, measured, and released back into the river, allowing the upstream journey to continue. Some of the eels were tagged with small microchips, known as passive integrated transponder tags. More recently, a semiautomated method has been used to count and measure eels using the eelway.[12]

Once the eelway was in place, I expected to capture young eels in the act of moving upstream, and ultimately to answer basic questions about the migration of yellow eels. How many eels are moving upstream? When are eels moving, such as time of day and time of year? Do river flow and the phase of the moon influence movements? Lastly, how big are the eels moving upstream?

From 2003 through 2018, the eelway at Millville passed 30,724 yellow eels. I do not know if all eels swimming up the lower Shenandoah use the Millville eelway. Some eels may have moved upstream by other means. It is possible, for example, that eels swim upstream through the canal system of the hydropower station. Based on anecdotal reports, eels may travel around dams on land during wet nights. Because of the other possible routes, the eelway count represents a minimum and not a total count of eels moving upstream. Also, I do not know if 30,724 eels represent a small or large number, because the numbers of eels swimming up the Shenandoah River are not known from 20, 50, 100, or 1,000 years ago.

What I do know is that the eels, ranging from three to ten years old, often move at night, climbing the eelway during all hours from dusk to dawn.[13] Most eels using the eelway were nearly 12 inches (30.5 cm) in length, usually ranging from 7.5 to 22 inches (19.1 to 55.9 cm).

Eels used the eelway from spring through fall. Upstream eel movements were largely linked to water temperature, river flow, and lunar phase.[14] Eels begin moving upstream during spring, when water temperatures reach 53.6°F to 59°F (12°C to 15°C). Following a rainstorm, an increase in river flow often coincided with a large number of eels using the eelway. High rainfall amounts in the Shenandoah watershed, as in most of the central Appalachians, happen in the spring and late fall. Summers and early fall, frequently hot and dry, have few rainstorms and low river flows. Eels also often move by the moon, not by the light of the full moon, but during dark nights near new moons. The largest numbers of eels used the ladder when spikes in river flows occurred at the same time as a new moon. Most of the largest catches at the Millville eel ladder were in the late fall during hurricane-induced floods. One memorable catch was observed near a new moon in November 2004, when over 2,000 eels used the eelway in one night, while the river raged with floodwaters from Hurricane Frances. Eels stop moving upstream during late fall, when water temperatures drop to about 50°F (10°C).

Our study showed that the Millville eelway serves its purpose: many eels used the eelway to pass upstream of the dam. You should consider, however, that eels were getting upstream before the eelway was in place. Likewise, eels are present upstream of four more hydroelectric dams on the Shenandoah River (figure 3.3). From Millville Dam, an eel travels another 45 miles (72.4 km) upstream, passing from West Virginia into Virginia waters, before

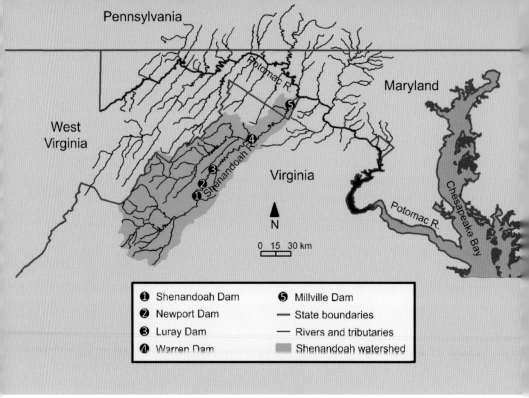

FIGURE 3.3. Five hydropower dams on the Shenandoah River.

reaching the Warren Dam. Continuing upstream for another 50 miles (80.5 km), an eel encounters a third dam, near Luray, Virginia, swims another 13 miles (20.9 km) to Newport Dam, and another 15 miles (24.1 km) upstream to the fifth dam at Shenandoah, Virginia.

We know that eels get upstream of dams at Millville, Warren, Luray, Newport, and Shenandoah because of the small and large eels that we have collected upstream. Plus, other biologists and anglers have caught many eels upstream. I have marked many of the eels captured at the Millville Dam eelway with passive integrated transponders (PIT tags). These small tags, enclosed in glass cylinders, are about the size of a Tic-Tac and are injected into an animal with a 12-gauge needle. The PIT tag emits a radio frequency that is detected as a unique number sequence by a PIT tag reader. This method of animal tagging is common for pets (dogs and cats) and livestock. Some people have even injected themselves with PIT tags. I have PIT-tagged American Eels with the hope of later recapturing them upstream, which would provide information on the timing by which they move upstream of the dams. The hard part is finding a tagged eel upstream—like looking for a needle in a haystack. Several have been recaptured, and one in particular was very informative. An 11.6-inch (29.5-cm) eel, PIT-tagged at Millville Dam on June 20, 2007, was recaptured at Millville Dam on October 2, 2008, at a length of 17.4

inches (44.2 cm), and then later recaptured 123 miles (198 km) upstream from the reservoir above Shenandoah Dam on October 14, 2009, when it was measured at 20.3 inches (51.6 cm). This was an interesting find because it demonstrated that while an eel may not move much at all, when it does decide to move, it can pass the Millville, Warren, Luray, Newport, and Shenandoah dams in a little over a year's time.

To understand the eel, one must not only study upstream movements of yellow eels, but also the downstream migration of adult silver eels. Most of my understanding of silver eels comes from the research efforts of my graduate students. One student studied the downstream migration of eels from the upper Shenandoah River watershed to the Chesapeake Bay.[15] During a three-year study (2007–9), she tagged and released 145 large eels in the upper Shenandoah River watershed, including 92 silver eels, 25 yellow eels, and 28 intermediate between silver and yellow.[16] All were scored as large eels, ranging in length from 28.4 to 40.1 inches (72.1–101.9 cm). The tags, about the size of a AA battery, emit unique radio frequencies that can be relocated and recorded with a radio receiver. The eels were captured in the upper Shenandoah River watershed, upstream of the dams, using an electrofishing boat—a boat that creates an electrical field in the surrounding water, which stuns fish and allows for their capture. After capture, eels were surgically implanted with radio tags and released back to the river.

We wanted to document the date and time that each eel passed a dam during its downstream journey, so several radio receivers were set up at each of the five dams—Shenandoah, Newport, Luray, Warren, and Millville. Also, radio receivers were set up farther downstream within the Washington Aqueduct on the lower Potomac River, as well as at Little Falls. After tagging and releasing the 145 eels, we waited for the eels to start their downstream journey, knowing that the signals from the radio tags would be recorded by the receivers as the eels passed the dams on their way downstream. The study answered several questions about how and when eels pass hydroelectric dams.

As expected, many of the radio-tagged eels began the journey downstream. By April of 2010, a total of 96 eels had begun their downstream journey, of which 81 were silver eels and 58 had passed the lower dam on the Shenandoah River at Millville. Water flow and moon phase influence downstream movements, where eels mostly moved during an increase in river flow or on dark nights near the new moon. During higher river flows, both the amount of water and the speed of water movement increase, which make it easier for eels to swim downstream. Also, given its nocturnal nature, an eel avoids light by swimming downstream during muddy water conditions of higher flows or during dark nights near new moons.

On each dam, strategically placed receivers recorded whether an eel passed over the spillway or passed using the raceway toward the turbines; sometimes the method of passage was not determined. For example, 84 of the radio-tagged eels passed the Luray Dam, of which 51 passed over the spillway, 28 used the raceway, and 5 were undetermined. Farther downstream, at Warren Dam, 63 eels passed the dam, with 41 swimming down the spillway, 19 using the raceway canal, and 3 being undetermined. Out of 58 eels passing the Millville Dam, 31 used the spillway, 26 used the raceway canal, and 1 was undetermined. Some of the eels that followed the flow into the canal raceways were killed by the turbines. Depending on the dam, the annual mortality rates ranged from 15% to 39%. But one must consider cumulative effects, because eels migrating downstream must pass multiple hydroelectric dams on the Shenandoah. Based on data for the three years of study, the estimated cumulative mortality rate was 53%.

In an effort to reduce turbine-related mortality of silver eels in the Shenandoah, the turbines are shut down at night from mid-September to mid-December. This management strategy is based on the expectation that most silver eels migrate downstream during fall and early winter. My student's study documented downstream movements past dams in every month except July. The shutdown period was effective at reducing eel mortality—the cumulative mortality rates of eels were 20% during the shutdown period versus 62% at other times.

If a little over half of the silver eels leaving the Shenandoah are killed by hydroelectric dams, then you may wonder how this impacts the overall population. Furthermore, given that silver eels migrate out of hundreds of rivers along the Atlantic Coast, you may wonder whether the loss of eels from one river impacts the overall population. Keep in mind that many rivers have hydroelectric dams, so the overall impact is cumulative. Something to consider is that silver eels migrating downstream are females with eggs. Interestingly, a large silver female has many eggs—not hundreds, but likely millions.[17] For most fishes, most eggs will not survive, most larvae will not survive, and most juveniles will not survive to a mature spawning adult. This is probably also true for eels, although no one knows the percentage of three million eggs that would survive to adulthood—one-tenth of a percent survival would equal 3,000, about one-tenth of the number of eels that we estimated swimming up the Shenandoah during a 15-year time period. It seems plausible that each female silver eel could make a valuable contribution to the population.

American Eels, with sinuous curves and long-distance migrations, are symbols of our rivers—the free-flowing connection between the central Appalachian highlands and the sea. Dams have removed the free-flowing wildness of our rivers, and are dangerous for eels. For most, it is difficult to discuss nature and the environment without placing value in terms of good or bad. I think that most people see the production of both electricity and American Eels as good. The tradeoff, I think, is in trying to sustain the two. The next time you flip on a light switch and sit in your reading chair, consider reading more about eels—a truly amazing and fascinating fish of the central Appalachians and beyond.

Smoke-Screen Foraging

In his Nobel lecture in December 1973, Niko Tinbergen spoke the opening remarks: "Many of us have been surprised at the unconventional decision of the Nobel Foundation to award this year's prize for Physiology or Medicine to three men who had until recently been regarded as mere animal watchers."[1] Tinbergen and his fellow ethologists—scientists involved with the comparative study of animal behavior—Karl von Frisch and Konrad Lorenz received the award for their discoveries of individual and social behavior patterns of animals. Tinbergen's work was on understanding the behavior patterns of sticklebacks, a group of fishes with species in North America and Eurasia. During his lecture, he went on to discuss how this "old method" of "watching and wondering" about behavior could contribute to the relief of human suffering caused by stress.[2]

Like Professor Tinbergen, I think that watching and wondering are wonderful ways to enjoy nature, satisfy curiosity, and advance science, particularly when learning about fishes. As a young naturalist, I was continuously watching and wondering. My summers were partially spent with my grandparents, who lived in the former coal camp of Raleigh, West Virginia. There I waded up and down Whitestick Creek, a heavily degraded stream that at times smelled like an open sewer. But I thought it was a natural paradise. My curiosity in nature had me wading and often crawling in the creek, chasing water snakes, crayfish, larval salamanders, and fishes.

Unfortunately, the curiosity of children is often lost with age, yet I am convinced that some of the watching and wondering of youthful years guides later decisions in life. For example, in Tinbergen's 1952 paper "The Curious Behavior of the Stickleback," he provided some background on what promoted his interest in this group:

> When I was a young lecturer in zoology at the University of Leyden
> 20 years ago [i.e., the early 1930s] I was asked to organize a laboratory
> course in animal behavior for undergraduates. In my quest for ani-
> mals that could be used for such a purpose, I remembered the stickle-
> backs I had been accustomed as a boy to catch in the ditches near my
> home and to raise in a backyard aquarium. It seemed that they might
> be ideal laboratory animals. They could be hauled in numbers out of
> almost every ditch; they were tame and hardy and small enough to
> thrive in a tank no larger than a hatbox.[3]

Tinbergen's youthful experiences clearly stayed with him in his later life.

Fish watching is not solely for scientists. It is a common human behavior, enjoyed by the young and old at heart. According to the *2017–2018 U.S. Pet Ownership and Demographics Sourcebook* of the American Veterinary Medical Association, about 10.5 million US households have pet fish.[4] A living room aquarium brings nature indoors. Research shows that people experience re-duced levels of stress and anxiety from watching aquarium fish.[5] An increasing number of dental offices have fish tanks (think of *Finding Nemo*), as patients benefit from reduced waiting room stress. I can only conclude that Tinbergen was right: watching and wondering about aquarium fish reduces stress and human suffering. Interestingly, few people go outdoors to watch fish. Here is something to try on a Saturday afternoon—stand on the shoreline or stream bank of a clear water lake or stream and use binoculars to watch fish. Feeling adventurous? Go snorkeling or diving for a relaxing fish-watching experience. Another approach is to use an underwater video camera. I have done all of these things either for recreational relaxation or as part of scientific research.

I have observed many fascinating fish behaviors during my favorite pastime of fishing. As an example, while ice fishing on Deep Creek Lake, a reservoir in western Maryland, my son and I witnessed an interesting feed-ing strategy for Northern Pike. We were targeting Yellow Perch using small plastic jigs tipped with waxworms. We were fishing in 16 feet (4.9 meters) of water, and the lake bottom was a smooth layer of sediment. Perch often roam these "mud flats" in search of food. We had an underwater camera set up so that we could see fish approach our lures, an effective fishing strategy and one that allows for observation of fish behavior. We were just talking about how long it had been since we caught our last Yellow Perch when a Northern Pike appeared on the video screen, approaching my son's lure but hesitating about six inches (15.2 cm) away. I expected the pike to lunge at the lure. Instead, it paddled the bottom sediment with an alternate rowing motion of its left and right pectoral fins. This fin-churning motion created a large cloud or plume

FIGURE 4.1. Chronological sequence of video frames depicting smoke-screen foraging behavior of a Northern Pike. (a) The Northern Pike approached the fishing lure (center), (b) initiated a sediment plume by rowing motion of its pectoral fins, (c) swam forward within the sediment plume, (d) completely concealed itself within the plume, (e) moved upward out of the plume toward the lure, and (f) engulfed the lure.

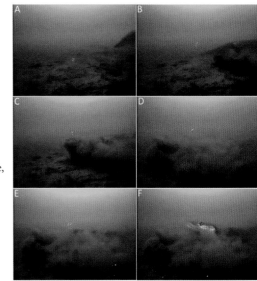

FIGURE 4.2. My son and his catch, a 27-inch (68.6-cm) Northern Pike, photographed inside our ice shanty. The excitement of observing this fish and its smoke-screen foraging behavior was followed by the fish becoming momentarily entangled in the camera cord. Fortunately, the Northern Pike was successfully brought up through the hole in the ice, photographed, and released back into the lake.

of sediment, which appeared to me to be something similar to a military maneuver known as a smoke screen. The pike, while completely hidden in the expanding sediment plume, swam forward and then upward to engulf the jig (figures 4.1, 4.2). This behavior differed from what I expected, as Northern Pike are known to attack prey with a "fast-start" burst of acceleration.[6] After watching this interesting pike behavior, I started wondering . . . Why would a Northern Pike, or any fish for that matter, use a "smoke screen" as a foraging strategy?

As someone who studies fishes, I have always been intrigued by the Northern Pike (figure 4.3). This fish has an elongate profile with its median fins located far back on the body. Its body shape is similar to that of several close relatives, including four found in the central Appalachians (Muskellunge, Chain Pickerel, Grass Pickerel, and Redfin Pickerel). In North America, the Northern Pike has a wide northern distribution, with the northern part of the central Appalachians as the southern extent of its native range, although populations have been widely introduced elsewhere to rivers, lakes,

FIGURE 4.3. Photograph of a 25-inch (63.5-cm) Northern Pike from a Central Appalachian reservoir.

and reservoirs for fishing opportunities. The Northern Pike is a premier predator perched at the apex of the trophic fish pyramid. Individuals can reach lengths of 50 inches (127 cm). With a mouth full of razor-sharp teeth, the Northern Pike has been given many colorful names, including gator, water wolf, and my personal favorite, sharptooth McGraw.

The prey-attacking tactic of the Northern Pike is a fascinating foraging foray. Pike often remain motionless and allow prey to come to them, a strategy known as lie-in-wait. Sometimes Northern Pike cruise along weedy edges or across mud flats in search of prey. The interesting part of foraging follows when prey is encountered, when the pike initiates its fast-start attack, a high-powered burst of acceleration from a resting position. This leads us back to Niko Tinbergen, who was involved with much of the initial research on feeding strategies of Northern Pike, as an indirect consequence of his stickleback research. Tinbergen was interested in how sticklebacks (the prey) reacted to Northern Pike (the predator). By watching and wondering, he discovered a lot of interesting things about how the Northern Pike attacks its prey. Tinbergen and his colleagues were the first to describe the S-start foraging of Northern Pike, a behavior now known in many predatory fishes.[7]

In the initial stages of attack, the Northern Pike forms its body into an *S* shape, with median fins erect.[8] The *S* shape results from simultaneous muscle activity—anteriorly on one side of the body and posteriorly on the opposite side. The anterior part of the body is less curved than the posterior part. The *S* shape is the starting point of attack, analogous to the three-point stance of a defensive lineman before exploding forward at the snap of the football. The Northern Pike achieves large-amplitude movements with its long and streamlined body, where the increased surface area of the erected posteriorly positioned median fins propels and thrusts the fish forward toward the prey.[9] The mouth is closed until just before reaching the prey.[10] The lunge of the Northern Pike is fast. In a study that looked at the acceleration rate of feeding Northern Pike, researchers found, on average, a total attack time of 0.13 seconds covering a distance of 8.2 inches (20.8 cm), which converts to about 5.2 feet per second (1.6 meters per second); a mean maximum velocity was estimated at 10.1 feet per second (3.1 meters per second).[11]

"A life subdued to its instrument," wrote Ted Hughes,[12] an eminent modern English poet known for his animal poetry. He was describing the Northern

Pike, and the line suggests that the pike "does what it does," that is, it has natural instincts that drive its behavior. As a researcher and angler, one thing I have learned about fishes is that they do not always "do what they do." Sometimes they do other things. Nature is not always predictable. Northern Pike are not always fast-start foragers. This is demonstrated by my video, where the Northern Pike did not forage with a fast start, but rather created a cloud of suspended silt, from which it emerged slowly to engulf the fishing lure.

The pike created and appeared to use the cloud of silt like a smoke screen, a strategy used in warfare for centuries. The basic premise is to create a cover of "smoke" that provides concealment.[13] The use of military smoke screens has changed with time. As it is lighter than air and difficult to breathe, smoke is not always an effective screen. During World War II, the US government obtained the "smoke screen" technology of Alonzo Patterson, a successful Louisiana rum runner during the Prohibition years who was responsible for several inventions, including advancements on speedboat designs. Patterson's smoke screen did not involve smoke but was based on what he called "fog oil," a mixture of three primary components: an oil base, an inorganic mineral salt, and a viscosity-inducing agent.[14] The fog oil was vaporized, creating a dense fog and a very effective "smoke screen."

Smoke-screen foraging seems like an effective strategy for aquatic predators, but I have found only a few mentions of it in the scientific literature. Mud-plume feeding is a foraging behavior used frequently by bottlenose dolphins in the Florida Keys.[15] A bottlenose dolphin creates a mud plume of linear or curvilinear shape in shallow water, while swimming and disturbing the bottom substrate at the leading edge of the plume. Then the dolphin repositions itself to lunge through the plume as a foraging behavior to capture prey. Similar behavior has been observed for estuarine dolphins along the eastern coast of Brazil.[16] Thus, there are examples of smoke-screen foraging in marine mammals, but other than our single observation of this behavior with Northern Pike, I know of none for foraging fishes.

I have repeatedly watched the video of my son's Northern Pike and its plume of sediment, wondering whether this is a feeding strategy or an incidental happening. Possibly the Northern Pike was just slowly progressing forward with use of its pectoral fins, and the sediment plume was an unintended consequence. A sample size of one does not provide very strong scientific inference. Maybe someone else will witness smoke-screen foraging in Northern Pike, providing further evidence. In the meantime, I will continue ice fishing, an occupation that I recommend for reducing stress and human suffering. Further, I will continue watching my underwater video camera while fishing, wondering whether I will see this interesting behavior a second time.

Karpfen

The Common Carp (*Cyprinus carpio*), with an appropriate common name, is commonplace in most places where it occurs, including the central Appalachians. Its connoisseurs are uncommon. In fact, most central Appalachians do not favor this exotic fish, as it is often considered a poor meal and an unwelcome inhabitant. The Common Carp, also known as the German Carp or European Carp, has been introduced widely around the world. I chose to title this chapter "Karpfen" based on the German common name. Its native range may include the Danube River,[1] which originates in Germany and passes through many European countries before reaching the Black Sea. The Common Carp was intentionally introduced to the central Appalachians in the late 1800s. Primarily, three varieties were introduced: Scale Carp, Mirror Carp, and Leather Carp (figure 5.1). The Scale Carp is currently the most common variety in our waterways. The Mirror and Leather Carps are domesticated varieties, rarely seen in the wild.

The history behind the scientific and common name of this species was summarized in 1995 by Eugene K. Balon: "The generic name of carp from Greek—*kyprinos* or *kyprianos*—was the name used by Aristotle (384–322 B.C.; 1862) and was probably derived from 'Kypris' (lat. *Cypria*), a secondary name of the goddess of love, Aphrodite—perhaps because the high fertility of carp and noisy mass spawning in shallows was known even then. Later the name was latinized to *Cyprinus,* probably by Pliny the Elder." The specific name of "carp" comes from the Celtic vernacular name for the fish and dates from the time when Celtic tribes inhabited the current eastern Austrian and Slovak-Hungarian shores of the River Danube. From the original *charpho, carfo,* and *charofo* the name gradually changed to the present *carpio*.[2]

FIGURE 5.1. Photograph of a 14-inch (365-mm) Common Carp (scale variety) from the Central Appalachians, and illustrations of Scale Carp (*top*), Mirror Carp (*center*), and Leather Carp (*bottom*) from Cole (1905), who described them as "drawings made for the Bureau of Fisheries from fish in its ponds in Washington soon after the introduction of the species into this country."

In the first edition of *The Compleat Angler,* Sir Isaak Walton wrote favorably of the character and hardiness of the nonnative carp in England:

> The Carp is a stately, a good, and a subtle fish, a fish that hath not (as it is said) been long in England, but said to be by one Mr. Mascall (a Gentleman then living at Plumsted in Sussex) brought into this Nation:

and for the better confirmation of this, you are to remember I told you that Gesner sayes, there is not a Pike in Spain, and that except the Eele, which lives longest out of the water, there is none that will endure more hardness, or live longer then a Carp will out of it, and so the report of his being brought out of a forrain Nation into this, is the more probable.[3]

In subsequent editions of *The Compleat Angler,* the previous lines were edited to elevate the carp's status to the "queen of rivers": "The Carp is the queen of rivers; a stately, a good, and a very subtil fish; a fish that hath not been long in England, but is now naturalized."[4] The *Compleat Angler* has been reprinted many times,[5] and the availability and popularity of this book from the mid-1600s to present has likely done more than any other source to positively influence people's opinions of the carp's character.

An understanding of how this species came to be widely distributed throughout the central Appalachians requires a look into the career of one of our native sons, Spencer Fullerton Baird (1823–87). An avid naturalist, Baird had a knack for collecting natural history specimens and a strong desire to become a museum curator. He was born in Reading, Pennsylvania, a railroad town near the eastern edge of the central Appalachians. As a young man, Baird lived in Carlisle, Pennsylvania, a small borough located in the Cumberland Valley of the Great Valley region of the central Appalachians. Baird began collecting specimens as early as age 15,[6] and corresponded often with the naturalist John James Audubon.[7] At age 13, Baird entered Dickinson College of Carlisle in 1836, graduated with a bachelor of arts degree in 1840, and became a professor of natural history there in 1845.[8]

Baird's dream of becoming a museum curator became reality in 1850, at age 27, when he was hired as the assistant secretary at the Smithsonian and first curator of the National Museum. In 1924, the naturalist Clinton Hart Merriam emphasized Baird's influence on the formation of the National Museum: "It is a matter of record that when 27 years old he had accumulated 3,500 skins of birds, upwards of 500 glass jars and numerous barrels, kegs and tin vessels of reptiles, 600 skulls and skeletons of North American vertebrates, a number of embryos 'in pickle,' and large collections of fossil bones from the bone caves of Pennsylvania and Virginia. This enormous collection when shipped to the Smithsonian in October, 1850, filled two railway cars and became the National Museum."[9]

In 1871, Baird was given additional responsibilities as he was appointed the commissioner of fish and fisheries of the US Fish Commission by President Ulysses S. Grant. As explained in 1883 by the ichthyologist George Brown Goode,

The resolution establishing the office of Commissioner of Fisheries required that the person to be appointed should be a civil officer of the Government, of proved scientific and practical acquaintance with the fishes of the coast, to serve without additional salary. The choice was thus practically limited to a single man, for whom in fact the office had been created. Professor Spencer F. Baird was appointed and entered at once upon his duties. Being himself an eminent man of science, for forty years in the front rank of biological investigation, the author of several hundred scientific memoirs, no one could realise more thoroughly the importance of a scientific foundation for the proposed work.[10]

As commissioner of fish and fisheries, Baird was responsible for the "introduction and multiplication of useful food fishes throughout the country."[11] In the late 1800s, the US population was increasing and the introduction of freshwater food fishes seemed like a reasonable option for increasing food supplies. In Baird's view (based on his understanding of carp in Europe), the German Carp was an excellent candidate for introduction, a species that could be raised in farm ponds for food. In 1874, he stated that "sufficient attention has not been paid in the United States to the introduction of the European carp as a food-fish, and yet it is quite safe to say that there is no other species that promises so great a return in limited waters."[12]

Europeans are fond of carp and have enjoyed it as table fare for centuries. Consider the supporting passage from 1882 by A. Woldt:

Not just recently, but for many centuries, does the fondness of our people exist for the savory, nutritious, golden carp, a dweller of the muddy and humus-rich grounds of our shallow waters. Its peaceful, quiet natural nature has stamped this fish with at least as great rights to a domestic animal as the bristle-bearing meat supplier of our rural population. Throughout the Middle Ages, carp has always been one of the most popular foods, even today, more than ever. It is among the most distinguished in all strata of the population, even in the preparation as "carp in beer" to a national court of the Germans. Its meat contains important nutrition, comparatively as much as that of the best part of cattle and is for spicy, spicy preparation very suitable; at the same time, it is free of the unpleasant taste of many other fish.[13]

In 1876, based on his understanding of carp in Europe, Baird provided the following list of carp qualities:

1. Fecundity and adaptability to the processes of artificial propagation.
2. Living largely on a vegetable diet.
3. Hardiness in all stages of growth.
4. Adaptability to conditions unfavorable to any equally palatable American fish and to very varied climates.
5. Rapid growth.
6. Harmlessness in its relations to other fishes.
7. Ability to populate waters to their greatest extent.
8. Good table qualities.[14]

The initial ponds for holding and rearing carp were put in place at Druid Hill Park in Baltimore, Maryland, and subsequently Babcock Lakes in the Monument lot of Washington, DC. Next, carp for the ponds were obtained from Germany. In 1905, the initial story of the German Carp in the United States was described in a nearly 120-page monograph by Leon Jacob Cole. Cole's carp interest was perhaps a career interlude, as he was generally known as a geneticist, ornithologist, and the father of American bird-banding. Cole wrote:

In the winter of 1876–77, Mr. Rudolph Hessel, in the interests of the Fish Commission, as an initial experiment shipped carp from Bremen to Baltimore, but, owing to a storm of unusual severity to which the vessel was exposed, all were lost on the way. He immediately returned to Europe, however, where, at Hochst, near Frankfurt, he procured another lot of fish. These he succeeded in bringing in safety to New York, and on May 26, 1877, they were placed in ponds in Druid Hill Park, Baltimore. This lot consisted of 345 fish, of which 227 were naked and mirror carp, and 118 were common scale carp. The ponds at Druid Hill Park not being sufficient for the proper care of the fish, Congress allowed use to be made of the Babcock Lakes in the Monument lot, in the city of Washington, and appropriated the sum of $5,000 to put these in proper condition. In the following spring these ponds were ready for the reception of the fish, and 65 leather carp and 48 scale carp were transferred to them from the Druid Hill Park ponds.[15]

In subsequent years, more ponds were created and additional carp were imported from Germany. As expected, the adult carp spawned successfully, and their offspring were distributed to individuals by an application process. Specifically, a person interested in receiving carp could complete an application of the US Fish Commission, then send the application through a US senator or representative to Professor S. F. Baird.[16] The Fish Commission distributed the carp for free, although the recipient paid for the cost of transportation.[17] For example, in 1882 more than 7,000 applications were submitted, and 5,758 applicants received a total of 143,696 fish.[18] The Fish Commission distributed carp to applicants within the US for 17 years (1880–96).

The carp program was initially considered by many to be extremely successful, as emphasized in 1884 by Ralph Stockman Tarr: "By introducing the German Carp to America a work of great economic importance was achieved, and the large number of carp-ponds in America shows the popularity of this new fish."[19] Carp were widely distributed within the United States, as documented by Cole: "As illustrating how thoroughly carp were disseminated throughout the United States in these early years of its introduction, the data for 1883 furnish an interesting example. In that year carp were sent into 298 of the 301 Congressional districts, representing 1,478 counties; in this way 260,000 carp were distributed, in lots of 20, to 9,872 applicants. The distributions continued large until about 1890, when they began to diminish, and were finally discontinued in 1897."[20]

Although the success of the program was measured initially by the number of applicants and the widespread distribution of carp, another measure of success was the acceptance of carp as a food fish. Interestingly, articles published in the *Bulletin of the U.S. Fish Commission* (edited by Baird or his colleagues) provided favorable reviews on carp as table fare. For example, a letter to Professor S. F. Baird from H. B. Davis of Macon, Georgia, published in the 1882 *Bulletin,* stated that "they are not full of bones, and do not taste of mud, as some would have us think, but, on the contrary, are very free from small bones, and are a most excellent table fish, to which several who have dined with me will testify."[21] A letter to S. F. Baird from Edward Thompson of St. Johnland, Suffolk County, New York, published in the 1884 *Bulletin,* stated: "The carp is the best fish I know of for workingmen and mechanics, who rarely lack an appetite, and who will always consider the fish good when they can get it. My personal opinion is that it is a very superior fish, and I will even go so far as to say that I prefer it to trout."[22]

In the 1883 *Bulletin,* editor of publications Chas. W. Smiley published an article titled "Answers to 118 Questions Relative to German Carp." Four of the questions were about carp as a food fish, and are presented below with answers (A.):

> What kind of a food-fish is carp?—A. Equal or superior to catfish, suckers, perch, and all our common native varieties. Many correspondents declare them equal to trout, bass, and shad, but this is not claimed for them by the Fish Commission.

At what age are carp suitable for table use?—A. When small the bones are troublesome, but the flavor is the same. When they weigh 4 pounds or more the bones can be easily removed.

What season of the year are carp fit for the table?—A. From October to May. During and for several months after spawning the flesh is soft. No fish is in good condition to eat for some time after its spawning time.

Do carp have many bones?—A. What fish does not? The flesh flakes off, however, very nicely from large carp.[23]

Also in the 1883 *Bulletin,* Smiley published an article titled "Notes on the Edible Qualities of German Carp and Hints about Cooking Them."[24] Smiley summarized the results of a questionnaire prepared and sent out in July 1883 by the US Fish Commission to individuals who had received young carp in 1879 and 1880.

> There have been received thus far over 600 replies, and from these have been copied verbatim everything which correspondents have said with reference to this subject. Over 350 had not yet tasted carp. Finding that there was an occasional remark of an uncomplimentary character, I inquired of Professor Baird with reference to publishing any adverse statements. In reply, he said: "Certainly it is not our policy to suppress honest criticism of the carp, and you are authorized to collate the testimony and publish both sides. No FISH IS FIT TO BE EATEN DURING OR IMMEDIATELY AFTER THE SPAWNING SEASON. Unless criticisms have been made of the fish during the late fall or early spring they do not affect the question."[25]

Smiley summarized 242 responses to the questionnaire, of which 204 were favorable and 38 were uncomplimentary or with some reservations.[26] Individual responses from those of central Appalachia are presented in appendix 2. Smiley was concerned about how people would view the lopsided results:

> I have spoken thus definitely concerning this material because when treated with exact impartiality the opposition to the food qualities of carp dwindles down into such utter insignificance that someone might easily suspect me of culling the material. This point is especially to be guarded, because it is so often considered praiseworthy to suppress criticism which is prompted by ignorance of facts and which

might injure a good cause. And I am very sure that every unfavorable opinion of carp herewith, . . . is based upon ignorance and forgetfulness of one of three facts:

I.   No fish should be eaten during or immediately after the spawning time.
II. The muddy taste of any fish can be largely removed by keeping the fish alive in a tub of pure water, changing it daily for a week.
III. Bad cooking will spoil the best of food.[27]

As noted, Baird listed eight qualities that he thought supported the carp's candidacy for introduction. Most of these qualities are unquestioned. Undoubtedly, female carp have high fecundity, that is, the propensity to produce large quantities of eggs. In 1878, Rudolph Hessel reported that a four- to five-pound female produced 400,000 to 500,000 eggs.[28] Carp spawn successfully in nature and in propagation ponds. Carp are certainly adaptable to many environmental conditions. Carp introduced to the United States were not constrained to ponds, but quickly escaped or were introduced into lakes, rivers, and creeks. Carp invaded the aquatic habitats of colder northern climates to warmer southern climates. As Gottschalk (1967) noted: "Their fecundity and ability to populate waters was demonstrated in Lake Erie. A few carp were stocked in western Lake Erie in 1883. By 1889, commercial fishermen took 3,633,679 pounds."[29]

There is no doubt about the hardiness of the Common Carp, a quality emphasized in 1653 by Izaak Walton. The species' hardiness was reflected in its ability to survive ship transport over long distances. Consider the following interesting account of Leather Carp shipped from the Washington, DC, ponds to Scotland, based on a December 6, 1881, letter to S. F. Baird:

DEAR PROFESSOR BAIRD: You will be pleased to hear that the 25 leather carp have been safely landed after a very stormy voyage. The gale was a very severe one, and on Wednesday, November 23, the wind blew with hurricane force, and we were obliged to "heave to" for twenty-two hours; the seas were very large indeed. We shipped one during the night which disabled the four seamen on watch; one had two ribs broken and another his head badly cut; the other two were lamed. The wheel-house was "stove in" and the galley bulged in; one boat was carried away on deck, breaking down the chimney-stack of the donkey-engine. Through all this storm the carp did well.[30]

The rapid growth rate of carp introduced into the United States was noted in 1883 by George Brown Goode: "They prove to be especially adapted to our waters, and in some localities they grow with surprising rapidity. A fish, 4 inches long, placed in the waters of Texas, was found to have increased to 20 ½ inches in eleven months, at which time it weighed 4 pounds 11 ounces."[31] Leon Cole further emphasized the Common Carp's rapid growth:

> In 1881 the Ohio State Fish Commission put 40 carp into the Maumee River (Ohio Fish Commission Report, 1882, p. 1435), and in May of the same year some were planted in Ten Mile Creek. These were 2 ½ inches long when liberated, and it is reported that in the following September and October a number were caught which would weigh 4 ½ to 5 pounds, while one had a weight of 8 pounds.[32]

Carp not only have rapid growth, but also attain large size. In the central Appalachians, the state weight records (as of 2018) for the Common Carp caught by hook and line are 54.9, 50, 47.5, 52, 49.3, and 47 pounds (24.9, 22.7, 21.5, 23.6, 22.3, and 21.3 kg) for the states of Kentucky, Ohio, Maryland, Pennsylvania, Virginia, and West Virginia, respectively.

Carp are omnivores, feeding on a wide selection of food items. Aquatic vegetation, zooplankton, and benthic invertebrates are all on the menu, but availability of food items varies seasonally. Baird's suggestion that carp live largely on a vegetable diet may be true during warmer seasons. Carp often focus their foraging efforts on sifting insect larvae from soft bottom habitats. Their benthic foraging uproots plants.[33] Carp can decimate aquatic vegetation, as noted from the carp's population explosion following introduction in western Lake Erie:

> Coincidentally, there was a major decline in aquatic vegetation, particularly *Vallisineria* [wild celery], *Sagitteria* [arrowhead] and *Zizania* [wild rice]. There was also a decline in the canvasback duck (*Aythya valisineria*) population in western Lake Erie. By 1900, most of the desirable aquatic vegetation was gone as were the canvasbacks. Similar reports were received by the U.S. Fish Commission from duck clubs throughout the United States.[34]

The decline of duck populations suggests that the introduced carp was not "harmless" to other species. Similarly, many people began to question Baird's claim that the Common Carp is harmless in its relation to other fishes. As previously addressed, Common Carp reduce aquatic vegetation, thus removing habitat for other fishes.[35] Bottom-rooting by a single carp creates a plume of sediment in the water column, and water bodies with high carp densities often become muddy or turbid.[36] Water turbidity reduces sunlight penetration, which results in decreased primary production. Carp also act as a predator and can reduce macroinvertebrate densities.[37] Thus, carp can negatively influence other fishes within a food web through bottom-up effects on nutrient availability and through predatory top-down effects.[38]

The collapse of the carp-introduction program corresponded with the growing perception of the carp's unsatisfactory flavor as a food fish and the potential of carp to harm native fish populations. Even during the carp-distribution program of 1880–96, people were starting to question the value of carp as a food fish, voicing their opinions to their government representatives. On January 4, 1884, Baird sent the Kentucky-native John Griffin Carlisle (a conservative pro-business Bourbon democrat and Speaker of the House of Representatives) a letter in support of the carp:

> There is naturally much difference of opinion as to the value of the carp as an article of food. No one who has at his command the choice fishes, such as salmon, trout, whitefish, mackerel, sheepshead, red snapper, &c., would be likely to attach a high value to the flesh of the carp. But in Germany and Austria it constitutes the principal article of consumption in the interior, and brings precisely the same price in the city markets as the native trout. In Berlin it brings about 25 cents per pound. Much, of course, depends upon the mode of cooking and the idiosyncrasies of the taster.[39]

In 1900, the assistant secretary and statistician of the Pennsylvania Fish Commission, W. E. Meehan, was asked if stocking Common Carp would provide forage to support a Smallmouth Bass fishery. Although there was little evidence in support of carp as spawn eaters, Meehan responded, "My antipathy to the carp is so great on account of its well known spawn eating habits and its other undesirable qualities, that I cannot myself endorse this suggestion."[40] In 1902, from the *Seventh Report of the Forest, Fish, and Game Commission of the State of New York*, Richard Cotchefer (general foreman of hatcheries) was clear on his commission's view of the Common Carp: "I regret to report that there is apparently a very large increase of carp in many waters of the State, and in many instances they have nearly ruined the fishing, principally by the condition in which they keep the water. They apparently multiply very much faster than any other fish, and it is to be regretted that they were ever introduced into the waters of this State."[41]

Willson (1898), in an attempt to find usefulness for the nonnative Common Carp, suggested raising carp as a food source for otter farming: "While carp have recently been pronounced by the fish commissioners of several States an unmitigated nuisance, compared with which all the plagues of Egypt were but a mild chastisement, the very objection to the aquatic stranger—that it multiplies like some miserable species of insect—only adds to its value as a food supply for otter farming."[42]

Spencer F. Baird, a central Appalachian son, was one of the most influential ichthyologists and one of the greatest naturalists of the history of the United States. He made substantial contributions to science in the fields of not only ichthyology but herpetology, mammalogy, and ornithology. In 1923, the renowned ichthyologist David Starr Jordan listed Baird as one of his most influential mentors and noted that Baird published 1,063 books, papers, and scientific notes.[43] Baird's direct or indirect influence on contemporary and future scientists was enormous. But as Jordan also noted in respect to Baird, "The introduction of the carp (*Cyprinus carpio*) into the United States was on the whole unfortunate."[44]

I agree that the introduction of carp into the waters of the United States was a blemish on Baird's record, but perhaps just a small blemish in an overall large legacy. In defense of Baird's role in introducing the Common Carp, Leon Cole stated: "It was not the intention of the Fish Commission to introduce the carp into waters that were already stocked with good native species, nor was it claimed that the carp was superior to the majority of our indigenous food fish. But it was believed that it could be successfully raised in many sections of our country not favorable to the growth of better fish."[45]

Unfortunately, carp escaped from their designated ponds, entered rivers and reservoirs, and expanded their range throughout the central Appalachians. I predict that many of the carping critics among us will one day find a favorable use for this fish. Perhaps, as central Appalachians, we will one day cull our culinary criticisms, tastefully transition our idiosyncrasies, and clear a place for the Common Carp on dinner tables or restaurant menus. Cooking with beer, butter, lard, or bacon grease may ease this transition. Regardless, in the end, we must endeavor to endure the carp, as Cole succinctly summarized: "And when we have decided whether the carp does more harm than good, we still have the real question before us. The essential problem is this: The carp is here, and here to stay; what are we going to do with it? How can we make the most of its good qualities and prevent it from doing damage? Even were such a course desirable, the extermination of the carp in our waters is out of the question."[46]

Hornyhead

We central Appalachians use colorful names for fishes. Many of these names are part of my fond childhood memories. I fished frequently during my younger years, catching red-eyes, spotted cats, mud cats, blue cats, bugle-mouths, and hornyheads (figure 6.1). Red-eye is another name for Rock Bass. Spotted cats and mud cats were names for juvenile and adult Flathead Catfish. Although there is a species in the central Appalachians known as the Blue Catfish, the blue cat of my youth was a blue-phase Channel Catfish. Young Channel Catfish are sometimes called fiddlers. A bugle-mouth is a Common Carp.

As a child in the central Appalachians, I had hours of fun fishing for hornyheads. It was as much fun to say "hornyhead" as it was to catch one. A young angler needed only to find a small creek and have a hook, line, and worm. I did not cook and eat hornyheads, although that has been common practice. The term "hornyhead," as an Appalachian appellation, is well known by many mountaineers, but few know much about them. Through the years I have often been asked various forms of the question "Exactly what the heck is a hornyhead, anyhow?"

Although there is a fish species of the central Appalachians with the common name Hornyhead Chub, the colloquially used "hornyhead" does not apply to a single species. The hornyheads include many species of minnows, all of which share the character of having small "horns" on the heads of adult males during springtime. Three genera of fishes with the largest "horns" in the central Appalachians are *Nocomis, Semotilus,* and *Campostoma* (figure 6.2). Depending on your stream drainage location, you may find hornyheads of the genus *Nocomis,* including the Hornyhead Chub, River Chub, Bull Chub, Bigmouth Chub, or Bluehead Chub (figure 6.3). Creek Chubs and Fallfish (genus *Semotilus*) also develop horns during the breeding season, as well as the Stoneroller Minnow (genus *Campostoma*).

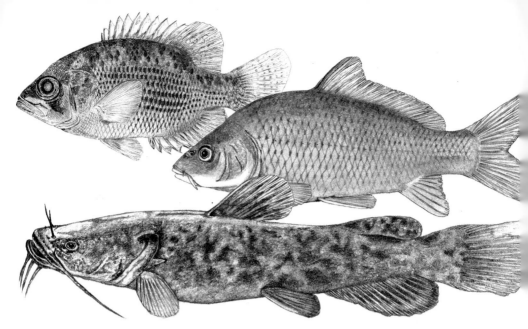

FIGURE 6.1. Three Central Appalachian fishes with interesting colloquial names; red-eye (*top*), bugle-mouth (*middle*), and spotted cat (*bottom*). Image of bugle-mouth adapted and redrawn from Jenkins and Burkhead (1994), courtesy of the American Fisheries Society.

FIGURE 6.2. Head profiles of hornyheads, including Bigmouth Chub (*top and middle*), Creek Chub (*bottom left*), and Stoneroller Minnow (*bottom right*).

FIGURE 6.3. A Bluehead Chub in its natural habitat with breeding tubercles visible on the head. Photograph by Ryan Hagerty, courtesy of the US Fish and Wildlife Service.

During the springtime, hornyheads seem to appear out of nowhere in our rivers and creeks. These small fishes, usually less than ten inches (25.4 cm) in length, are often caught with hook and line from trout streams, or by seine or minnow trap from creeks. Many Appalachian anglers seek out hornyheads for use as baitfish, as a hook baited with a three- to four-inch (7.6–10.2 cm) hornyhead will often produce a nice catch. Most anglers do not realize that hornyheads represent many species of minnow, or that the "horns" are seasonal. Many of our hornyhead minnows spawn during springtime, but the horns may develop during winter and persist through early summer, depending on the species.

Hornyhead presence in our streams during springtime is puzzling to some people, whereas others perceive these fishes as prized possessions. Robert E. Jenkins and Noel M. Burkhead recounted an interesting conversation about the appearance and disappearance of hornyheads in Virginia: "To one gentleman in southwestern Virginia, its disappearance in early summer is a mystery of nature. He asked if we knew where the 'hornyhead' went. We informed him that they don't go anywhere, that males are always present, but after spawning some die and others shed their 'horns' just like a buck deer loses antlers. He shook his head and said, 'That's not right.'"[1]

The Hornyhead Fish Festival is an annual celebration in Newborn, Georgia. This festival, near the southern edge of Appalachia, includes a fishing tournament. In 2018, a ten-year-old angler won the tournament with a hornyhead that was 7.5 inches (19.1 cm) long with ten horns. In the southern Appalachian region of the Great Smoky Mountains, hornyheads (i.e., Stoneroller Minnows) are targeted by anglers for sport and food.[2]

In scientific terms, the "horns" on the head of hornyheads have been referred to as multicellular horny tubercles,[3] and more commonly as pearl organs, nuptial tubercles, or breeding tubercles.[4] The term "pearl organ" has been in use for a long time; the German anatomist Friedrich Maurer used the term *Perlorgane* in 1895.[5] Martin L. Wiley and Bruce B. Collette commented that "in some species, tubercles have a pearly color which makes the use of the term 'pearl organ' appropriate, but many other fishes wherein the term has been applied, lack such coloration."[6] Perhaps the term "breeding tubercle" is the most informative, as it accurately reflects that these tubercles (small protuberances) are present during the spawning season.

Breeding tubercles are capped by keratin, a structural protein. Keratin is not commonly found in fishes, although it is common to other animals; examples include reptile scales,[7] bird feathers,[8] and mammalian hair.[9] In addition to large breeding tubercles on the heads of hornyheads, smaller keratinized tubercles are found elsewhere on the body, including fins.[10] In addition to large head tubercles, some members of the genus *Nocomis* also have smaller tubercles on their pectoral fins and on the lateral sides of the body.[11] In *Semotilus,* large head tubercles are accompanied by small tubercles on the operculum (gill cover), the pectoral fins, and the posterior body.[12] In *Campostoma,* large breeding tubercles are found on the head, but smaller tubercles form on the back and posterior sides of the body, as well as on the dorsal and pectoral fin rays (figure 6.4).[13] My hope is that your inner naturalist is now inquisitive about the functional significance of breeding tubercles. What is the purpose of these horny structures?

One explanation, supported by observational studies, is that hornyheads use head tubercles as weapons to defend their nests from other males. Mark H. Sabaj described the aggressive nest-defense behavior of *Nocomis* males:

> An intruding male similar in size to the nestbuilder elicited intra-specfic agonistic behaviors characteristic of *Nocomis* species (e.g. head butts, parallel and circle swims). Head butts occurred as the attending male drove his tuberculate head into either the head or body of the intruding male. Aligned eye to eye two males might parallel each other's motion, locking cephalic tubercles while moving upstream. When

FIGURE 6.4. Adult male Stoneroller Minnow with breeding tubercles visible on the head, back, and side of body.

performing circle swims, males swim head to tail in a tight circle for as long as 10 seconds (Maurakis et al. 1991). Contests ended when the attending male expelled the intruder from the vicinity of the nest.[14]

In 1908, Jacob Reighard recounted aggressive encounters of breeding male Creek Chubs, which at the time were commonly called "Horned Dace":

Frequently other male dace approach the nest. If these are smaller than the nest builder they are pursued and then invariably flee. Such small males are distinguished from the larger ones by the presence of a dark lateral stripe. If the male dace that approaches the nest is of the same size as the nest occupant a battle frequently ensues. The two strike at each other with their heads in apparent efforts to inflict wounds with the sharp pearl organs. They often struggle together fiercely in these encounters, but neither fish appears to suffer any injury. The sole result seems to be to produce temporary discomfort in the fish that is hooked, and this usually results in the departure of the intruder.[15]

Rudolph J. Miller described the aggressive behavior of tuberculated male Stoneroller Minnows:

Large male *Campostoma* spent much time driving intruders from the nest. They usually use two behavior patterns, the swing and the butt. Both movements are aggressive and may be considered as different intensities of a direct frontal attack. This attack was seldom seen in its entirety. Basically it consisted of a movement of the body directly toward another stoneroller with the anterior part of the body curved so the tubercles of the head and dorsum were directed against him. At times this movement resulted in a butt, called a "body check" by Smith (1935: 159). A butt may injure the attacked fish, but usually the intruder fled before being struck. At its highest intensity the attack continued as a chase for a short distance. The swing may represent the intention movement for the frontal attack. It consisted of a turn and short charge toward the intruder and sometimes was accompanied by a tilting component which oriented the tubercles toward the opponent.[16]

In another explanation for the functional significance of these knotty-heads, Michael R. Ross suggested that breeding tubercles on the heads of males may serve as "display structures signaling status in a social system based on male rank."[17] Thus, breeding tubercles may play a role in male dominance. In many minnows, dominant males obtain and successfully defend the highest-quality breeding territories, which often result in higher reproductive success than that of their less-imposing male colleagues. Males of larger size are decidedly dominant. Well-developed breeding tubercles may also determine male dominance, but larger-sized males often have larger size and numbers of well-developed head tubercles. Thus, there is a tight linkage between larger body size and larger weapon size, both possible determinants of a male's ability to attract females and defend breeding territories. This linkage confounds our understanding of the contributions of male body size versus breeding tubercles to male dominance.

Further, Ross made an interesting comparison of horned minnows with that of horned mammals: "In horned mammals (Geist, 1966) and in creek chubs, the development of display has included the development of head accessories. These accessories can inflict wounds, but also have the potential to signal the probable outcome of overt aggression, and thus avoid fighting. Such a behavioral system could lead to large savings of energy (Geist, 1966), which could be diverted to physiological maintenance and/or an increase in the investment of energy into successful reproduction."[18]

Perhaps another explanation for tubercle function is apparent at the forefront of the spawning act. Tubercles placed on the pectoral fin and posterior body probably aid males of some hornyhead species in holding the female during spawning, a behavior known as a "spawning clasp." In 1903, Reighard suggested that these breeding tubercles "serve chiefly to roughen the skin and enable the male to retain his hold of the female during the brief spawning act."[19] Subsequently, in 1908, Reighard presented his explanation for the function of breeding tubercles from a study of the Creek Chub (a.k.a. Horned Dace, figure 6.5):

FIGURE 6.5. A female (*vertical*) and male (*curved*) Creek Chub in the spawning act. The figure is adapted from Reighard (1908), who used the following caption: "Male and female horned dace during the spawning act. On the male, which is the fish with the body curved, are shown above the eye and in line with the nostril, four spine-like pearl organs and below these two smaller spines. Small organs are seen on the operculum and dorsal surface of the pectoral fin and on the caudal edges of the scales on the tail."

A close examination of a breeding male of the dace is necessary to show the means by which he retains his brief hold of the female in spite of the slipperiness of the skin of both fishes. If he be so examined, it is found that those parts of his body which are in contact with the body of the female during the embrace are beset with minute, sharp pearl organs, and are thereby rendered rough, like a piece of sandpaper. The opercular region, covered with close-set pearl organs, has a shagreenlike feel. The sides of the body and tail from the caudal edge of the dorsal fin backward are provided with minute organs which occur in rows along the slightly everted edges of the scales and roughen the surface over which they are found. Finally the upper surfaces of the pectoral fins are provided with close-set organs of moderate size which form rows along the fin rays. By means of these organs the male, whose body would otherwise be smooth and slippery, is enabled to make effective his brief hold of the female.[20]

Nocomis chubs have a similar spawning clasp to that illustrated by Reighard for the Creek Chub.[21] The male spawning clasps of some species of Nocomis are likely aided by small tubercles on pectoral fins, or small tubercles on the lateral sides of the body.[22] Interestingly, breeding male Stoneroller Minnows have pectoral fin and posterior body tubercles, but do not clasp females like Nocomis or Semotilus during spawning. Miller and William R. McGuire found similar results from their studies on the spawning behavior of Stoneroller Minnows,[23] where the latter described the spawning act as follows: "As a female moves into the spawning pit, one to five males align themselves laterally to the female. She will then be turned so that her body is perpendicular to the current as the males press her into the substrate with their pectoral girdles and fins."[24]

Although I have grouped the hornyheads as three genera (Nocomis, Semotilus, and Campostoma), males of many other minnow species of the central Appalachians also have breeding tubercles. Some consider Pimephales minnows as members of the hornyhead club, which are represented in the central Appalachians by Bluntnose Minnow, Fathead Minnow, and Bullhead Minnow. Several other examples of minnow genera with male breeding tubercles are Clinostomus, Cyprinella, and Luxilus (figure 6.6).

FIGURE 6.6. Examples of breeding tubercles on the heads of several other male minnows of the Central Appalachians: (a) Bluntnose Minnow (Pimephales notatus), (b) Rosyside Dace (Clinostomus funduloides), (c) Steelcolor Shiner (Cyprinella whipplei), and (d) Common Shiner (Luxilus cornutus).

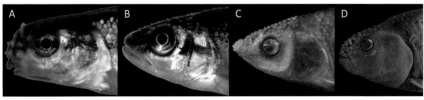

Now you know that the colloquial hornyheads of the central Appalachians do not represent a single species, as many of our minnows have males with breeding tubercles. Head and body tubercles of hornyheads likely serve various functions relative to spawning. Hornyheads caught by anglers are often species in the genus *Nocomis, Semotilus,* or *Campostoma,* although *Luxilus* species are also sometimes caught by trout anglers. Hopefully, this chapter has honed your knowledge about central Appalachian "hornyheads," a fun word for a fascinating group of fishes.

Nocomis Nests

The last publication of Jacob Ellsworth Reighard (1861–1942), "The Breeding Habits of the River Chub," was issued posthumously in 1943.[1] Based on data collected in 1904, 1905, 1915, and 1926, the paper was published with final editing by colleagues at the University of Michigan. Reighard was an American zoologist and ichthyologist, but he was also an old-school naturalist, as shown by his detailed field notes of the River Chub's breeding habits. The early naturalists often entered a state of nirvana during narration of field notes, which were later formatted as manuscripts and published. I enjoy reading the writings of early naturalists, as the excitement of their observations comes through in their prose. I have a few colleagues that have these wonderful tendencies, but unfortunately, the number of scientists focused on natural history has faded in recent years.[2]

Many naturalists, myself included, tend to observe one species at a time. However, inquisitive naturalists are also rewarded by observations on how two or more species interact, or how one species influences others. In some cases, habitat-altering actions of one species can have a large influence on other species. A well-known example from the central Appalachians is the construction of dams by Beaver (*Castor canadensis*). These stick dams alter stream ecosystems,[3] creating deeper and slower-moving aquatic habitat and thus directly impacting other aquatic species.[4] Animals that alter habitat, creating habitat used by other species, have been termed keystone modifiers,[5] or ecosystem engineers.[6]

In the central Appalachians, the River Chub (*Nocomis micropogon*) serves as an excellent example of an ecosystem engineer (figure 7.1). The River Chub is a large minnow. Adults range from 4.0 to 8.5 inches (10.2–21.6 cm) in length. Breeding males have head tubercles and often develop a huge

FIGURE 7.1. A male River Chub with head tubercles.

hump on their head, known as a nuptial crest. Because of the head tubercles, anglers regularly refer to these individuals as "hornyheads" (see chapter 6, "Hornyhead"). Most important for this chapter, mature male River Chubs are ecosystem engineers, as they alter stream habitat by building mound-nests out of pieces of gravel. But how could a gravel mound of a male River Chub affect the individuals of other species?

In the following description of how River Chubs build gravel mounds, I include prose from Reighard for an old-school flair. Reighard was an expert in methods for observing habits and life histories of fishes in nature, winning a prize of $100 in gold for a paper on the subject presented at the Fourth International Fishery Congress in 1908.[7] In "The Breeding Habits of the River Chub," Reighard detailed how he got close to a nest before making observations:

> Cautious approach was often rewarded by a sight of the fish at work. If they were not present a motionless wait would bring them back to their tasks. These waits often ran from five minutes to half an hour, but a longer wait was usually fruitless. After the fish had been busy for a while I often crept nearer on hands and knees and again stopped for a time, with my head lifted only high enough to enable me to see the nest. So by several stages I advanced stealthily until I was on the edge of the bank, and the fish were little, if at all, disturbed. I avoided all quick movements of hand or head, and did not permit my garments to flutter in the wind. After a time on the bank I often moved toward the nest slowly, with frequent stops, until I stood over it. A water glass (glass-bottomed container) that was sometimes used for closer observation did not disturb the fish for more than a few minutes. It is in this manner that one makes his first approach to a nest, but later approaches, it was found, could be more rapid.[8]

Reighard observed that a male first selects a suitable nest site. He then tries to remove pieces of gravel from several locations until he finds a spot where gravel pieces can be easily removed. Once the River Chub decides on a site, a first step in nest building is to excavate a pit. "He seizes any projecting part of a stone that he can get into his mouth at times jerking it from side to side to loosen it, and then turns partly on one side, lifts the stone, and swims away with it protruding from his mouth. If a stone is too large to be readily dislodged by lifting, the fish may put his opened mouth against it and slide it along the bottom."[9] The stones are removed and deposited primarily in areas that Reighard called "the dumps," which are located on each side of the pit, forming gourd-shaped areas with a wider section on the downstream end. The widths of gravel pieces removed from the pit often range from four inches down to a quarter inch (10.2–0.64 cm), but smaller sizes of gravel and sand-sized substrates are also removed. Reighard explained: "The stones that the male removes from the nest may be twice as large as its head and hence may project prominently from its mouth during transportation, but occasionally when it seems to be carrying nothing at all it may spit out sand or gravel. During this act its head is lowered, the mouth is opened, and finally the contents are suddenly discharged."[10]

Once completed, the somewhat circular or saucer-shaped pit is 12–15 inches (30.5–38.1 cm) across, with a central depth of 3–6 inches (7.6–15.2 cm). The male River Chub's next step is to build a spawning mound, one piece of gravel at a time. During the transition from pit to mound, some pieces of gravel are removed from the pit and some are added, until eventually the male is focused on just adding pieces of gravel. The male drops gravel pieces into the pit, leveling it with the stream bottom, thus creating a platform. He piles the pieces of gravel onto the platform to form a dome-shaped mound. The male often swims as far as eight feet (2.4 meters) upstream of the nest site to find suitable pieces of gravel, as downstream transport is easier. Suitable pieces of gravel from "the dumps" adjacent to the nest site are also added to the pile. According to Reighard, the presence of gravid females encourages the male to work faster: "Although the movements of females about a male engaged in the construction of a nest may interrupt his work, their advent upon the scene produces certain typical changes in the pace and character of the nest building activities."[11]

The male River Chub often takes several days to complete the gravel mound. He works during daylight, and often is most active in midafternoon. If environmental changes occur during nest construction, such as colder water temperatures or elevated stream flows, then nest completion may require a longer period of time. Reighard described the nest construction time under typical conditions as follows:

In 55 minutes, one male carried in enough stones to cover the pit bottom completely, and at the end of 1 hour and 25 minutes the pit was level-full. In 2 hours and 25 minutes the stones were heaped 2 inches [5.1 cm] above the plane of the stream bottom. By the end of the following day this male had accumulated enough stones to make a domelike pile 30 inches [76.2 cm] across and with a maximum height of about 3 inches [7.6 cm] above the bottom of the creek. Two days later the pile had grown to a diameter of 33 inches [83.8 cm] and a height estimated to be 6 inches [15.2 cm].[12]

Reading Reighard's paper, I asked, how many pieces of gravel does a River Chub need to carry to complete the mound? This number varies by nest, but Reighard estimated one nest to have 7,050 pieces of gravel. The average gravel size of this nest was 1.0 inch by 1.5 inch by 1.0 inch (2.5 by 3.8 by 2.5 cm), but some pieces of gravel were as large as 1.5 by 3.25 inches (3.8 by 8.3 cm) wide and 0.5–2.5 inches (1.3–6.4 cm) thick.

Reighard also estimated the distance that a male River Chub travels when building a mound:

In transporting a stone to the nest a fish often travels an estimated 12 feet [3.7 meters] each way and sometimes 20 feet [6.1 meters], that is, 24 to 40 feet [7.3–12.2 meters] for the round trip. If we conservatively estimate an average distance of 12 feet [3.7 meters] for round trip, we have probably made sufficient allowance for the relatively few trips on which two or more small stones are brought back. At 12 feet [3.7 meters] per trip the fish would have traveled 84,600 feet, or a little over 16 miles [25.7 kilometers], to accumulate his stone pile.[13]

Of course, the male River Chub's ultimate goal is to mate with females. Reighard noted that mating often occurs during the construction of the mound:

From time to time while the male is building his pile ripe females approach. The male then often stops his building activities, lowers his head very rapidly, and excavates a shallow spawning trough about the width of his body and about one half to two thirds his length. He does this by tossing or carrying stones from some place on the pile. The long axis of this minor excavation coincides with the current, and it is deeper at its upstream end. I have seen three such trenches made in ninety minutes by the same male. While hollowing out this little trough he stops at short intervals and lies over it with his head slightly elevated and pointed upstream; he then spreads his pectorals and for a fraction of a second rapidly vibrates the part of the body in the region of the vent, thereby rubbing his vent over the stones. This tremor may aid in cleaning the groove or may be an act of self-excitation preliminary to spawning. While the trough is being made or when it is completed a female usually enters and lies in it to spawn, as subsequently described. Spawning completed, the male returns with increased activity to his stone carrying and drops stones into the trough so rapidly that it is soon obliterated.[14]

Reighard not only described the reproductive and mound-building behavior of the River Chub, but also documented the use of the River Chub's nest by other fish species, including Central Stoneroller (*Campostoma anomalum*), Common Shiner (*Luxilus cornutus,* previously *Notropis cornutus*), and Rosyface Shiner (*Notropis rubellus*): "While the male is building his nest a horde of fifty to two hundred fish may intrude at one time, to spawn on it or seek eggs to eat. He does not try to eject them so consistently as do other minnows. He often moves undisturbed back and forth through the throng. The associated species give way little or much according to the size of the river chub, but only long enough to let him pass. Usually he behaves as though they were not there."[15]

Further, from Reighard's field notes dated May 4, 1905: "On the upstream edge of the nest are poised some half-dozen large males of *Notropis cornutus,* and below these hang many small males and females of the same species, perhaps 25 in all. Along with them are four or five females of *Nocomis* and many bright-red *Notropis rubellus* about three inches [7.6 cm] long."[16]

In my observations of *Nocomis* nests, I have noted several other nest associates (i.e., individuals that spawn on the nest of another species),[17] including Rosyside Dace, Striped Shiner, Southern Redbelly Dace, Mountain Redbelly Dace, and Longnose Dace (figure 7.2). Another interesting twist to this tale is that all fishes that frequent a River Chub's nest are not spawning. Some eat eggs of both the River Chub and nest associates. Egg predators documented at *Nocomis* mounds include Bluntnose Minnow, Northern Hogsucker, Johnny Darter, Greenside Darter, and Rainbow Darter.[18]

You may find it interesting that the male River Chub does not attempt to prevent other species of minnows from spawning on his nest. Possibly the more interesting part of this story, however, is that the River Chub actually benefits from having the nest associates. Nest associates with large body size, such as male *Luxilus* shiners, often chase off intruders from the gravel mound, such as egg predators. The aggressive behavior of some of the nest associates contributes to protecting their own eggs, but in doing so may also be protecting the eggs of the River Chub. Despite the protective

FIGURE 7.2. Gravel mounds constructed by *Nocomis* chubs, including an active nest (*top*) with nest associates (Rosyside Dace) and a postspawn nest (*bottom*) covered by silt.

efforts of nest associates, egg predators are successful at eating eggs from the gravel mound. However, the River Chub benefits from a "dilution effect,"[19] where the presence of eggs of both River Chub and nest associates reduces the probability of predation on River Chub eggs. River Chubs spawn during construction of the mound, so eggs are often not near the surface. If nest associates spawn on the perimeter or near the surface of the gravel mound, then egg predators approaching the nest from the top, sides, or downstream will first encounter eggs of the nest associates. By this "selfish herd" effect,[20] River Chubs reduce their risk of egg predation by centripetal positioning of their eggs relative to the egg locations of the nest associates.

In turn, nest associates benefit by spawning on the River Chub nest. In some streams, spawning gravel is sparse, and the River Chub creates spawning habitat for other species by aggregating pieces of gravel into a single location. In streams with plentiful gravel substrates, the nest associates are free to spawn elsewhere, but seem to prefer to congregate and spawn on the River Chub's gravel mound. Possibly, this preference results from the placement and physical characteristics of the River Chub mound. This may be related to the interstitial spaces among adjacent pieces of gravel, oxygenated waters, and lack of siltation at River Chub mounds, conditions that may not be maintained elsewhere during the spawning season. The presence of the River Chub may be more important than just the physical characteristics of the nest, as artificial mounds have been shown not to attract nest associates.[21] The nest associates also likely benefit from reduced egg predation by the dilution effect, as described previously.

Interestingly, the River Chub and its nest associates have mutual benefit from their nest-sharing relationship. This mutualistic relationship can be
thought of in terms of a biological barter.[22] The River Chub obtains a service
from the nest associates, as egg survival is increased by aggressive behavior
of some nest associates, and by the dilution and selfish herd effects. The nest
associates obtain a service from the River Chub, as a gravel spawning area is
provided, and the probability of egg survival is increased by the dilution effect.

Although I have covered the basics of this mutualistic relationship between the River Chub and its nest associates, the full ecological story is likely
to have more layers of complexity. The breeding and mound-building behavior of River Chubs are of extreme ecological interest, and many researchers
since Reighard's publication have added to our understanding of this process.[23] In the central Appalachians, the River Chub is not the only species
that constructs nests of gravel. Four other species of *Nocomis* chubs are
also present: Bigmouth Chub, Bull Chub, Bluehead Chub, and Hornyhead
Chub.[24] The Fallfish (genus *Semotilus*) also builds gravel spawning mounds,[25]
and its close relative, the Creek Chub (see chapter 8, "Spots and Dots") constructs a gravel nest known as a "pit-ridge."[26] The Tonguetied Minnow and
Cutlip minnow build nests with pebble-sized gravels.[27] As further studies
continue on gravel spawning minnows, we can expect further insights into
the mutualistic relationships between hosts and their nest associates.

I often marvel at the River Chub's work ethic in building gravel spawning mounds and the associated mating success. When considering the interactions of the River Chub and nest associates, it makes me appreciate the
complexity of the natural history of central Appalachian fishes. Just recently,
I found myself working at my home office and staring out the window at a
backyard pile of large stones. Presumably, these stones have been in the yard
for over 100 years, since construction of the house. They have made it easier
for me to relate to the extreme effort that River Chub's exert during mound-
building, since it seems that each new yard project requires me to move the
stone pile from one location to another. Unfortunately, my wife never seems
to be very impressed with the stones, or in my ability to move them.

Spots and Dots

Interesting stories are often linked with interesting people. Consider Samuel Latham Mitchill (1764–1831), a prodigious polymath of the late eighteenth and early nineteenth centuries. Often known for his Napoleonic blue coat, buff-colored vest, and shoes with buckles, he was a man who wore many hats: author, chemist, college professor, husband, journal editor, lawyer, legislator, naturalist, poet, physician, surveyor, US congressman, and US senator.[1] Mitchill was addressed as the "delphic oracle" by New York governor De-Witt Clinton,[2] and labeled as "the Congressional Library" by Congressman John Randolph of Roanoke.[3] President Thomas Jefferson called Mitchill "the Congressional dictionary,"[4] often inviting him to dinner and social events.[5] Like a keg of knowledge, "tap the doctor at any time, and he will flow" was a common description of Mitchill's character.[6]

Mitchill's grasp of the natural sciences was vast, including botany, chemistry, geography, geology, medicine, mineralogy, paleontology, and zoology. And he was fond of fishes! On New Year's Day of 1814, Mitchill published *Report, in Part, on the Fishes of New York*, with descriptions of many new species.[7] Subsequently, he described the new species *Cyprinus atromaculatus* in a volume of the *American Monthly Magazine and Critical Review*.[8] *Atromaculatus* is Latin for "black spotted." This species was later assigned to the genus *Semotilus*, a name proposed by one of Mitchill's colleagues, Constantine Samuel Rafinesque. *Semotilus* refers to the "black spot" on the dorsal fin, with *sema* for "banner (dorsal fin)" and *telia* for "spotted." This species currently has the scientific name *Semotilus atromaculatus*, but you may know it as the Creek Chub (figure 8.1). Currently, the Creek Chub is widely distributed across Appalachia. As you might expect from the name, Creek Chubs are common in creeks (small streams that you can jump across), but they are also found in larger streams.

FIGURE 8.1. Photograph of a Creek Chub. This fish is about three inches (7.6 cm) in length.

Creek Chubs have many interesting behaviors, two of which are shoaling and schooling. Shoaling is similar to schooling, in that both include a group of individuals. A shoal is a gathering of closely positioned individuals, where body positions of many individuals are oriented in different directions. A school is formed by individuals with synchronized swimming and body positions oriented in the same direction.[9] Many explanations exist for shoaling and schooling.[10] By shoaling, individuals may be more successful at finding food.[11] Food is generally not uniformly distributed in aquatic habitats, but rather often occurs in separate patches. A group of individuals will likely encounter a food patch at a higher rate than would a single individual. In Creek Chubs, shoaling and schooling behaviors are likely linked to predator avoidance.[12] For example, predator avoidance is more likely for individuals centered within the shoal or school, relative to those individuals positioned on the periphery.[13]

Creek Chubs also have interesting reproductive behaviors.[14] Just prior to spawning, males select nesting sites in gravel runs or often in areas at the base of pools with moderate water-current velocities. Nest-building nuptial males compete with other males for the best spawning site. Males are ornamented with head tubercles (see chapter 6, "Hornyhead"), which are essentially weapons for sparring with male competitors. Larger males usually win the battle. The male initiates a nest by moving sand and gravel (by mouth), creating a nearly circular depression on the stream bed. Spawning occurs at the upstream end of the depression, where the male fertilizes eggs from the female and then covers those eggs with particles of sand and gravel from the downstream side. Many females mate with a single male over several days, so the male is repeatedly moving material from the downstream to the upstream side to cover fertilized eggs. This type of nest has been termed "pit-ridge," where the pit is the nest depression and the gravel ridge is formed by the repeated covering of eggs (figure 8.2).

FIGURE 8.2. A pit-ridge nest. The male Creek Chub is oriented upstream. Adapted from Reighard (1910).

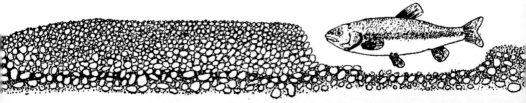

FIGURE 8.3. A breeding male Creek Chub, approximately seven inches (17.8 cm) in length.

Spawning involves female choice, where the female chooses to mate with a nesting male. It is unknown why a female finds a particular nesting male attractive, but it appears that females frequently select larger males who are often behaviorally dominant. Smaller, subdominant males are less successful at spawning. The dominance hierarchy may result from sparring, where larger males use their size and head tubercles to acquire and defend prime nesting sites. Also, female choice may involve male coloration; spawning males often display a reddish or orange stripe at the dorsal fin base, and pink on cheeks and parts of the pectoral, pelvic, and anal fins (figure 8.3).

Mitchill likely did not have specimens with spawning coloration, otherwise he may have derived the species name from its color. Instead, Mitchill chose the name *atromaculatus* because his study specimen was speckled with black dots: *atro* is from the Latin word *ater* for "black" and *maculosa* for "spotted." One of the distinguishing characters of a Creek Chub is a distinct black spot at the base of the anterior rays of the dorsal fin (figure 8.1). Many have assumed that Rafinesque's *Semotilus* and Mitchill's *atromaculatus* make reference to this dorsal fin spot. Mitchill did not, however, mention a black spot on the dorsal fin in the species description, but rather discussed black dots as follows: "back, sides, belly, and fins irregularly marked by black dots, consisting of a soft or viscous matter, capable of being detached by the point of a knife without lacerating the skin."[15] Apparently, Mitchill incorrectly thought that this body covering of black dots was a descriptive characteristic of the fish. In the central Appalachians, individuals of Creek Chubs, as well as many species of fishes, are often covered with these black dots, an interesting story that unfolds with answers to a few questions. First, why are Creek Chubs commonly covered with these spots? Second, what effect do they have on Creek Chubs?

You've probably noticed that I've called the Creek Chub's markings both spots and dots. Spots and dots are not always equivalent, a point raised by the savant Theodor Seuss Geisel (i.e., Dr. Seuss) in his story on "How to Tell a Klotz from a Glotz."[16] In reference to Mitchill's Creek Chub, black spots are the same as black dots, though in my experience the word "spot" is used more often than "dot" when referring to patterns of pigmentation on fishes. A black spot on the anterior rays at the base of the dorsal fin, in combination

FIGURE 8.4. A Creek Chub with black spot disease.

with other characters, is useful for distinguishing Creek Chubs from other species. Black spots covering the body of a Creek Chub are atypical, but are common in some streams, owing to external black cysts on the body and fins that contain trematodes. A trematode is a parasitic flatworm, often called a fluke. Scientists sometimes refer to this condition, where a fish is covered in these circular black cysts, as "black spot disease" (figure 8.4).

So it seems that the Creek Chub with black dots that Mitchill described was parasitized by black spot trematodes. In my research studies within the central Appalachians, I commonly see fish with black spot disease, although I have never considered the condition as a negative, but rather an interesting ecological story—a life cycle of a parasitic trematode that involves a complex relationship among a bird, a snail, and a fish.

Mitchill, in the early 1800s, did not know about the black spot trematode. In 1877, a small note on this condition, paraphrased from the British journal *English Mechanic and World of Science*,[17] was published in the *Annual Record of Science and Industry* for 1876, titled "Cause of the Black Spots on the Scales of Fish":

> The abnormal occurrence of black spots or specks upon the scales or external surface of fishes has frequently been observed, and quite often mistaken for regular coloration. Dr. Fatio, of Geneva, however, has been investigating some of these cases, and finds that in nearly all of them a small parasitic worm occupies the centre of this spot, and is easily observable by the microscope. This is inclosed in two cysts, with a peculiar liquid between, the inner being oval and transparent, and the outer round, with thick fibrous walls, outside of which is the mass of star-shaped pigment cells. The further stages of this worm have not yet been worked out, although it is quite probable that when the fish is devoured by its predaceous neighbors, this enters into another stage of the alternations of generation which have become so familiar of late years to investigators.[18]

Since then, researchers have pieced together the fascinating life cycle.[19] Many taxa of trematodes cause black spot disease, some of which are still undescribed species. The basic life cycle typically involves a piscivorous (fish-eating) bird as the final host, an aquatic snail as the first intermediate host, and a fish species as the second intermediate host. In the field of parasitology (the study of parasites), the term "host" is applied to an individual that harbors a larval or adult stage of a parasite. Depending on the trematode species, the final host may include a kingfisher or various species of waterfowl and herons. The intermediate hosts include various species of aquatic snails and fishes.

The adult black spot trematode lives in the gut of the bird host. Although adult trematodes are hermaphroditic (both male and female), it apparently takes two trematodes to tango. The trematode eggs are shed with the feces of the bird host, hatch after entering the water, and release the free-living larval stage, known as miracidia. A miracidium penetrates the integument of the first intermediate snail host, ultimately resulting in another larval stage, known as a cercaria. The cercaria (about 2 mm long) emerges from the snail's body many days after miracidial penetration. The free-living cercaria, which has a tail (used for swimming) and an oral sucker (suction mouth), penetrates the integument of a fish, often at the base of a fin. During penetration, the cercaria sheds its tail and transforms into another stage, known as a metacercaria. The metacercaria embeds in the fish host tissue, and within a few hours it secretes a cyst wall around itself. The fish deposits black pigment around the encysted metacercariae, resulting in the black spots characteristic of the disease. The bird host ingests a black-spot-infected fish, and the metacercariae are liberated from their cysts. The freed metacercarial larvae mature following attachment to the intestinal mucosa of the bird host, thus restarting this complex life cycle.

Coincidentally, as a beginning graduate student, I considered a thesis on black spot trematodes. I conducted a pilot study on Wills Creek, a stream with headwaters in Pennsylvania near the towns of Fairhope, Hyndman, Meyersdale, and lower sections passing by the Maryland towns of Ellerslie, Corriganville, and LaVale, before confluencing with the Potomac River in Cumberland. A total of 17 fish species were collected, during the study of which black spot disease was present on 15: Central Stoneroller, Fallfish, Spotfin Shiner, Striped Shiner, Rosyface Shiner, Spottail Shiner, Mimic Shiner, Bluntnose Minnow, Northern Hogsucker, Bluegill Sunfish, Rock Bass, Smallmouth Bass, Greenside Darter, Tessellated Darter, and Fantail Darter. Of the 15 infected species, not all individuals had the disease, suggesting that some individuals may not be susceptible to infection. Black spot disease was not present on Blue Ridge Sculpin or Potomac Sculpin. Species without scales, such as sculpins and catfish, are often not infected by this parasite. I have also not seen an infection in American Eels, a species with smooth skin and embedded scales. The intensity and prevalence of black spot trematodes within a fish assemblage will depend on the densities of bird and snail hosts, as well as the density and habitat preference of susceptible fish species.[20]

Research has shown that black spot disease may affect shoaling behavior in some species. In one study, Western Mosquitofish with black spot infections were less likely to join shoals compared to uninfected individuals.[21] Further, this study found that uninfected individuals preferred to associate with each other, but did not associate with infected individuals. Moreover, infected individuals did not desire to associate with other infected individuals. Although one individual can easily see black spots on another individual, a parasitized individual may not realize that it is infected. Thus, the end result is that infected individuals may be considered as nondesirable shoalmates or as outcasts by their conspecifics.

A study of black spot disease and shoaling in Banded Killifish included parasitized shoals (all individuals with blackspot disease), unparasitized shoals (no individuals with black spot disease), and mixed shoals (both parasitized and unparasitized individuals).[22] As expected, Banded Killifish discriminated against shoals with all parasitized individuals. However, individuals did not show a preference when given a choice of joining an unparasitized shoal or a mixed shoal. In the mixed shoal, parasitized individuals were often found on the periphery. It is not clear if peripheral positions were by choice or if the shoalmates forced the individuals to the outside. Predation risk increases for a fish in a peripheral position in the shoal, particularly when the fish is in the front.[23] Possibly, a fish may choose to join a mixed shoal, because its predation risk is reduced by the presence of parasitized peripherals.

External parasites have another interesting twist, a possible impact on fish mating success. In Creek Chubs, as in many fishes, females and males do not just hook up willy-nilly. The female makes a choice to mate with or to not mate with a particular male. Females likely select male mating partners based on physical appearance. Attractiveness in male Creek Chubs may indicate an individual that is in excellent health. The appearance of excellent health also suggests that a male may have "good genes," including parasite-resistant genes.[24] Studies on sticklebacks have suggested that males with parasite resistance may attain brighter breeding colors, thus female choice may involve bright coloration as an indicator of genes for parasite resistance.[25] In

FIGURE 8.5. A Creek Chub heavily infected with black spot disease. Note the eroded fin condition.

my experience, I have yet to see a dominant breeding male Creek Chub with black spot disease. In Creek Chubs, an infected male covered in black spots may just not be an attractive mate choice for a female. It is also likely that heavily infected males may have reduced stamina, unable to perform territorial defense and nest building. To my eyes, slightly or moderately infected Creek Chubs appear normal other than the black spots, but heavily infected individuals often have eroded fins (figure 8.5)

Researching this story piqued my interest in the subject of parasitology, but I am not sure how much I want to share about it. I expect that Samuel Latham Mitchill would have been able to pull off an engaging conversation about host-parasite relationships, although not likely at a dinner party. I do find that parasites, particularly fish-parasite relationships, are interesting, but I know that the topic makes many people uncomfortable. Because of this sensitivity, I will not follow with other chapters on parasite relationships, but interesting stories could be told about the largest parasite that affects humans, the fish tapeworm, or about smaller nasties such as liver and lung flukes, whipworms, roundworms, and screwworms, just to name a few.

Anomalous Anatomy

Nature's biodiversity in central Appalachia is reflected not only in the number of fish species, but in the species with unique and specialized morphologies and behaviors. The morphology of a fish refers to the form and structure of external and internal body parts. Consider the Central Stoneroller Minnow (*Campostoma anomalum*), hereafter called the Stoneroller (figure 9.1), a species with some unique and specialized external and internal body parts. The Stoneroller occurs throughout the central Appalachians, where it is most commonly found in rocky riffles and runs of creeks to medium-sized rivers.

FIGURE 9.1 Photographs of Stonerollers, including a breeding male (*top*) and a gravid female (*bottom*).

The Stoneroller has a near-ventrally positioned mouth and is a benthic (stream bottom) forager. For fishes, the position of the mouth often influences foraging location. Mouth positions in fishes are generally described along a gradient from ventral to dorsal positions, where categories include inferior, subterminal, terminal, supraterminal, and superior. For the benthic Stoneroller, the position of the ventrally located mouth could be characterized as nearly inferior or extremely subterminal. Fishes with terminal mouths have foraging flexibility, as they can forage on or near the bottom, midcolumn, or water's surface. Fishes with supraterminal or superior mouths often feed near the water's surface.

The Stoneroller has a unique mouth morphology. When viewed from underneath (ventrally), the mouth has the shape of a horseshoe. A specialized character of its mouth is a hard ridge along the lower jaw (figure 9.2). For the inquisitive naturalist, specialized body structures raise questions and demand explanation—of what relevance is the hard ridge on the lower jaw?

Furthermore, the Stoneroller has a uniquely shaped intestine. Edward Drinker Cope, a nineteenth-century naturalist who covered paleontology, herpetology, and ichthyology, provided some details on the internal morphology of Stonerollers in his 1869 publication *Synopsis of the Cyprinidae of Pennsylvania:* "In the remarkable genus *Campostoma* the canal extends to between eight and nine times the length, and, like that of other vegetable-feeders, is usually found occupied by the ingesta for a considerable part of its length."[1] In context, Cope was commenting on two interesting observations. First, the Stoneroller has an extremely long intestine, the length of which ranges from eight to nine times the length of its body. Second, the Stoneroller's intestine is often nearly full, suggesting that the species frequently forages for food.

FIGURE 9.2. Ventral view of the head of a female Stoneroller, illustrating the horseshoe-shaped mouth and the hard cartilaginous ridge of the lower lip.

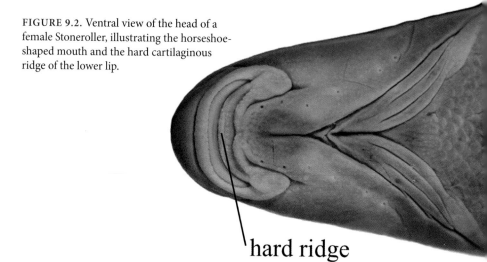

hard ridge

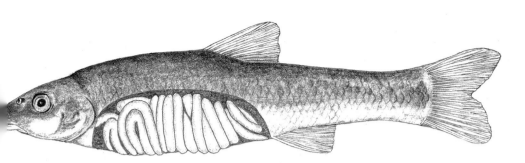

FIGURE 9.3. Lateral view of the abdominal cavity of a female Stoneroller, depicting the long, coiled intestine. Adapted and redrawn from Jenkins and Burkhead (1994), courtesy of the American Fisheries Society.

Further, Cope noted that the Stoneroller's long intestine was coiled around the swim bladder (figure 9.3). This explains how such a long intestine can fit inside the relatively small abdominal cavity of the Stoneroller. He also noted that the swim bladder of the benthic Stoneroller is reduced in size relative to some pelagic minnow species. In minnows, the swim bladder is a gas-filled sac used for buoyancy control, a useful organ for pelagic (water column) species, but perhaps less necessary for the benthic Stoneroller. Cope wrote:

> Indeed, in no other genus than *Campostoma* have I observed the latter organ [swim bladder] involved in vertical coils of the intestines, and separated from contact with the abdominal walls.
>
> [The swim bladder is] much diminished in size, not extending behind the anterior two-thirds of the abdominal cavity. This is appropriate to the abode of the species, near the bottom, where they obtain their favorite vegetable food. In the more numerous species of carnivorous habit, that rise to seize insects on the surface, the bladder extends throughout the length of the abdomen.[2]

When considering innards, most people tend to think in terms of human anatomy—a stomach connected to a small intestine followed by a large intestine. The internal anatomy of the Stoneroller Minnow was examined in separate studies by Walter C. Kraatz (1924) and Mary Dora Rogick (1931) of Ohio State University.[3] The Stoneroller does not have a true stomach or large intestine. Thus, in addition to the hard ridge of the lower jaw, the long and unique coiling morphology of the intestine is a second specialized structure of the Stoneroller. But why does the Stoneroller have such a long coiled intestine?

Stonerollers can be easily caught with a seine, a rectangular meshed net that has a lead line, a float line, and a brail (pole) on each end. Using a seine requires wading in the stream, which is no walk in the park. The benthic (bottom) substrates of streams often include rocks of various sizes, ranging from sand-sized particles to gravels, cobbles, and large boulders. When stream wading, your feet get wedged between larger rocks. Stepping on top of stream rocks can also be treacherous, as surfaces are often slippery. As a young naturalist, I first experienced slippery rocks while chasing crayfishes in creeks. At the time, I did not give the slightest thought to the slipperiness of the rocks, as I much enjoyed slipping, falling, and the subsequent soaking by the stream water. Today I still enjoy chasing crayfishes and fishes among the slippery stream rocks, but soakings are no longer appreciated.

The slipperiness of stream rocks is important to Stonerollers, a relationship that requires further explanation. The surfaces of stream rocks are often slippery owing to algae. Algal growth on rocks is quite common, particularly in sunlight-exposed sections of streams. The Stoneroller, often considered an algivore, is a herbivorous fish that eats algae from benthic substrate surfaces of streams. Algae on stream substrates is not something that most people give much consideration, but it can be quite diverse and include many organisms, such as different kinds of diatoms (single-celled algae structured with silica), green algae, and cyanobacteria (a type of bacteria often called blue-green algae). Detritus (dead organic matter) and small insect larvae are also present. Altogether, this slippery organic covering on the surface of stream substrates is sometimes called the periphyton. Thus, when feeding on the slippery surfaces of rocks, Stonerollers are likely consuming several different kinds of food.

Matthews, Power, and Stewart described how Stonerollers remove algae from rock surfaces, a foraging tactic that provides some insight into the importance and function of the hard ridge of the lower jaw: "When feeding *Campostoma* remove attached algae from such surfaces, they usually do so with behaviors we describe as 'swiping' (the head is thrust downward and sideways, with the lower jaw scraping algae from the rock in the sideways motion); 'shoveling' (pushing the lower jaw against the surface and swimming vigorously forward); or 'nipping' (small, rapid bites with the body at 45° angle to the substrate)."[4]

Gut length of herbivorous fishes generally exceeds that of carnivores. Presumably, this difference in gut length corresponds with the lower-quality food of herbivorous fishes. It seems like the herbivorous *Campostoma* would need to consume a lot more food than a carnivore to meet its protein requirements. Often food availability influences a fish's diet. Food is often available and abundant for *Campostoma*. If algae are absent or in low abundance, then Stonerollers may eat other food items.

In the heavily forested headwater streams of the central Appalachians, much of the energy input comes from terrestrial sources, such as autumn leaves falling into the stream.[5] Stonerollers are not generally abundant in these small, forested streams. As we go downstream in the watershed, small creeks come together. The confluence of streams contributes to a wider main waterway, where shade from riparian trees is restricted to stream margins. In these larger waterways, the energy input shifts from terrestrial toward aquatic sources. Specifically, energy inputs begin to include contributions from aquatic autotrophs, organisms that obtain energy from sunlight. Algae are often autotrophic, capturing the sunlight, providing a base for primary production, and providing a food source for Stonerollers.

Stonerollers are not strictly or seasonally herbivorous, as individuals are known to occasionally eat aquatic insect larvae, either selectively or possibly inadvertently while scraping the periphyton. During the breeding season, Stonerollers may require more protein in their diets, including invertebrates. For example, Stonerollers are commonly caught on a worm-baited hook by anglers during this time (see chapter 6, "Hornyhead"). Edward Cope noted this, stating, "The *Campostoma* is a fish of comparatively slow movements, and readily takes a hook baited with a worm."[6] Cope also observed *Campostoma* foraging on fish eggs dislodged from the stream bottom by Mountain Redbelly Dace (*Chrosomus oreas*), stating that "the *Campostoma,* too, of pale tints, and painted fins, swimming in pairs on the bottom, would gather with ease whatever the stream carried from the burrowing *Chrosomus.*"[7] In some unusual circumstances, *Campostoma* will consume the fry of other fishes.[8]

When algae are abundant, nutrient requirements likely find Stonerollers foraging on benthic algae frequently throughout the day. Fowler and Taber found that young Stonerollers can eat up to 27% of their body weight during a day.[9] Gut content is largely evacuated at night, a time when Stonerollers are not foraging.

Recently, physiologists have become interested in how gut microbiota affect human health. A similar question could be asked about herbivorous fishes, such as Stonerollers. Fish gut microbiota could aid digestion, particularly by enzyme-producing microbiota.[10] In 2017, Hannah Grice reported a large number of bacterial phyla and genera in guts of a stoneroller species from Georgia, but concluded that "*Campostoma* may not rely on symbiotic bacteria to aid in digestion."[11]

In 2009, Donovan P. German suggested that *Campostoma* guts act as plug-flow reactors,[12] where (1) digesta moves in one direction through the intestinal tract, (2) digestion does not rely on fermentation,[13] (3) digestive enzymes are most active in the foregut, and (4) absorption of nutrients mostly occurs in the mid to posterior part of the intestine. It seems that the function of the Stoneroller's intestine is somewhat simple. Stonerollers must eat a lot of algae to get their required amounts of nutrients.

Stoneroller populations may influence or be regulated by algal production. To add further light to the Stoneroller's story, I refer back to an earlier statement of foreshadow: "Algal growth on rocks is quite common, particularly in sunlight-exposed sections of streams." The population density of Stoneroller Minnows in a stream is influenced by the amount of algal growth. In the central Appalachians, streams with dense canopy cover of trees have less light reaching the stream during the warmer periods of the year. Streams with heavy canopy cover often have lower population densities of Stoneroller. Stonerollers are often most abundant in larger creeks to medium-sized rivers, where algae are abundant and the canopy cover of riparian trees is less effective at reducing sunlight exposure. Also, in some of our central Appalachian streams, sewage or agricultural runoff increases stream nutrients, supporting an increase in algal growth regardless of canopy cover.

Stoneroller populations may be influenced by predators or competitors. Stonerollers, because of their small size, are prey for larger predatory fishes. Stonerollers sometimes find themselves right in the middle of an interesting two-directional trophic cascade.[14] The first option, known as a top-down effect, occurs when predators reduce the population size of Stonerollers, allowing an increase in densities of benthic algae. In some cases, the presence of predators reduces daily foraging time of Stonerollers, thus lowering the losses of algae. The second option is where algal production is limited, resulting in lower densities of Stonerollers and other herbivores, which in turn leads to fewer predators (a bottom-up effect). By eating algal-grazing invertebrates, invertivorous fishes may increase the amount of algae available for Stonerollers. In this complicated scenario, Hargrave et al. found that the invertivorous Orangethroat Darter (*Etheostoma spectabile*) reduced densities of algal-grazing invertebrates, thus increased periphyton biomass by a top-down effect, which ultimately resulted in a bottom-up effect where the increase in periphyton enhanced growth of Stonerollers.[15]

Benthic algae in our central Appalachian streams, although important to populations of Stonerollers, have a wider importance at an ecosystem level. Without primary production by the aquatic autotrophs, aquatic animals such as insects, crayfishes, and fishes would be totally dependent on detritus or food entering the stream from terrestrial sources, and this would not be an adequate replacement for supporting the stream ecosystem. The importance of algae as a food source was established by William P. Coffman, Kenneth W. Cummins, and John C. Wuycheck, who examined the diets of annelid worms, aquatic insects, snails, crayfishes, and a fish (total of 75 species) in Linesville Creek, Pennsylvania.[16] Specifically, Coffman, Cummins, and Wuycheck found that 74 of the 75 species ingested algae, and 50 of the 75 had a diet of at least 50% algae.[17] Thus, Stonerollers have some competition for algal resources, particularly in streams with limited algal production.

You now have some insight into the specialized form and function of the Stoneroller's hard-ridged lower lip and long coiled intestine. Additionally, you now have an appreciation for algae. The hard ridge of the lower lip allows Stonerollers to graze (scrape) algae off of hard substrates on the stream bottom. The long intestine aids in the digestion of algae, a low-quality food. I encourage you to visit a stream and observe the Stoneroller's fascinating algae-foraging behavior. In the immortal words of Edward Drinker Cope, "The observation of fishes in their native haunts, brings much of beauty, as well as of interest, to the eye of the naturalist."[18] Whether by snorkeling or using binoculars from the stream bank, observing benthic Stonerollers foraging on algae would be a wonderful way for you to spend an afternoon. Give it some thought, as there is likely to be a lot more to learn about Stonerollers. Further observations could bring much of beauty to the eye of a central Appalachian naturalist, particularly for someone seeking a better understanding of this species and its trophic role in streams.

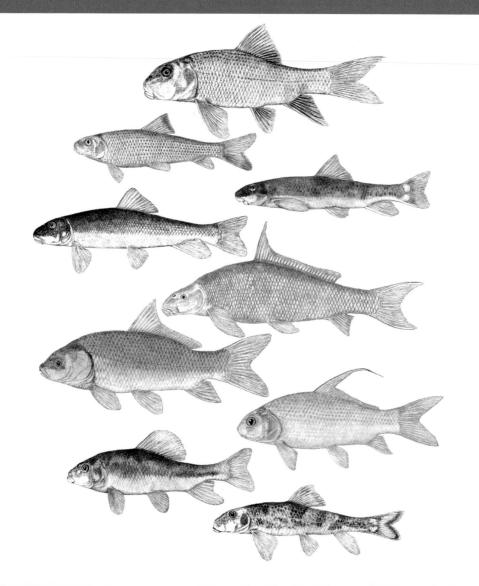

Sucker Savvy

Attitudes about central Appalachia's fishes are diverse. Some fishes are highly favored, a few have apathetic approval, while others are just straight-up scorned. I think it's often a fish's behavior that influences people's perceptions. For example, some fishes live at or near the water's surface, others mingle in midcolumn, and then there are those who dwell on the bottom. Bottom dwellers, also known as benthic fishes, are often among the least liked and least understood. Yet many central Appalachian fishes are bottom dwellers. Thus, to appreciate fully the diversity of central Appalachian fishes, we must be able to find beauty in benthic fishes. But why are the eyes of many beholders unable to find this beauty?

The *Merriam-Webster Dictionary* has three definitions for "bottom-feeder": (1) "a fish that feeds at the bottom," (2) "one that is of the lowest status or rank," and (3) "an opportunist who seeks quick profit usually at the expense of others or from their misfortune."[1] The first meaning was in use long before the second or third, so somewhere along the way, someone decided that low-status people and opportunists were analogous with bottom-feeding fishes!

But not all bottom dwellers are belittled. Two generally well-liked and diverse families of bottom fishes are Percidae (darters, Yellow Perch, Sauger, and Walleye) and Ictaluridae (catfishes). Who could find fault with a bottom-feeding Fantail Darter (see chapter 23, "Darter of Darters") or a tiny Tippecanoe Darter (see chapter 22, "Tippecanoe Is Tiny Too")? Who would not find beauty in brightly colored breeding males of the benthic Candy Darter (see chapter 17, "Charismatic Candy")? Walleye, Yellow Perch, and catfishes are often bottom dwellers, and are quite popular as sport fishes. For some anglers, their preference is primarily for eating them more than catching them with hook and line. Fillets of Walleye, Yellow Perch, and Channel Catfish, when coated with cornmeal and fried in corn oil, are classic central Appalachian cuisine.

Then there are those bottom dwellers of the family Catostomidae. Additional to the "bottom feeder" label, public perception further frowns upon these bottom fishes, in part because of their unflattering common name—sucker. The Catostomids have been labeled as rough fish, trash fish, and junk fish.[2] If there is one fish family at the bottom of the barrel of public perception, then it is the "poor" suckers—poor based on public perception, but otherwise phenomenal in all other regards. Perhaps it takes an ichthyologist to savvy suckers, but I hope this story will innervate your inner naturalist, increasing your awareness of the presence and diversity of suckers in our central Appalachian streams.

There's a sucker born every spring in our central Appalachian streams! Actually, there are lots of suckers hatched every spring, representing many species and contributing to the amazing aquatic diversity of the central Appalachians. Some sucker species are beautifully colored, some are large and some are small, some live in large rivers and others in smaller streams (figure 10.1). Some are distributed widely, whereas others have narrow ranges. The genera of native suckers of the central Appalachians have interesting common names, such as Redhorse and Jumprock Suckers (genus *Moxostoma*), Spotted Suckers (*Minytrema*), Torrent Suckers (*Thoburnia*), White Suckers (*Catostomus*), Blue Suckers (*Cycleptus*), and Buffalo (*Ictiobus*). Other names are both interesting and perhaps less appealing to public opinion, such as Carpsuckers (*Carpiodes*), Chubsuckers (*Erimyzon*), and Hogsuckers (*Hypentelium*).

Suckers are often incorrectly assumed to be simple and sedentary, not requiring clean stream habitats and moving just short distances upstream or downstream during their lifetime. In a review of spawning behavior in suckers, Lawrence M. Page and Carol E. Johnston dispelled these myths by summarizing some similarities among species: "All species spawn in the spring over coarse substrate, usually clean rubble or gravel, less often sand or vegetation. Generally, stream inhabiting populations spawn in shallow, usually fast-running water, and lake-inhabiting populations spawn along shallow shorelines. Most swim upstream to spawn."[3]

In addition to the use of clean substrate and upstream spawning migrations, the act of spawning itself is quite interesting in suckers. Consider the following captivating account of "trio spawning" by Page and Johnston:

> Probably the most notable characteristic of spawning in suckers is the intimate involvement of at least two males and one female in a

FIGURE 10.1. A Highfin Carpsucker from the Ohio River (*top*) and a Torrent Sucker from a small headwater stream of the upper James River drainage (*bottom*).

spawning act. Typically, a female swims into an aggregation of males, stops, and comes to rest on the bottom, facing upstream. Two males approach from opposite sides of the female, spread their fins, and press against her sides. During the spawning press, tubercles (if present) on the caudal peduncle and the anal and caudal fins of the males are pressed against the female (Reighard 1920, Jenkins & Jenkins 1980). Members of the spawning trio arch their backs, vibrate rapidly, and release eggs and sperm. During vibration, all three fishes rise anteriorly while digging into the substrate with their anal and caudal fins. This action dislodges coarse materials, and the eggs are buried as they are fertilized.[4]

Our suckers' upstream spawning migrations are generally not a point of interest, but perhaps they should be. In other regions, like the Pacific Northwest, migrating species of salmon capture attention, as these fishes promote angling, support subsistence fisheries, and boost economies. The financial value of salmon is difficult to measure in dollars, particularly when you consider one of their important contributions—the upstream transport and addition of nutrients which subsidize aquatic and terrestrial ecosystems.[5] In regions with salmon streams, nutrient subsidies are well studied, resulting partly from the post-spawn carcasses of dead salmon,[6] but also from nitrogen and phosphorus excreted by live fish.[7] Although less understood, upstream migrations of suckers can also result in nutrient subsidies. The spawning suckers do not typically die, so carcasses are not the subsidy source. Instead, spawning suckers fertilize the stream with nitrogen and phosphorus by excretion.[8] The influx of these nutrients has a bottom-up effect. The nutrients result in an increase in algal densities, which results in increased growth of macroinvertebrates, which in turn increases food availability for insectivorous fishes.

Suckers are completely entangled in the food webs of their stream communities. Young suckers provide a forage base for nearly all piscivorous fishes, and adult suckers are eaten by larger predatory fishes and birds. Suckers are mostly bottom feeders, and most eat macroinvertebrates (especially larval aquatic insects), along with some detritus and algae. As noted by R. E. Jenkins and N. M. Burkhead, "Suckers are able to sort and eject unwanted material via the mouth and opercular openings. Some materials are sensed by the taste-bud-laden lips; judging from gut contents, often only desired food is engulfed."[9]

I am not sure why bottom feeding in suckers is often ill-considered. It is almost as if some people assume that suckers are swimming along slurping sludge from stream bottoms. Interestingly, the native Brook Trout of the central Appalachians commonly feeds on aquatic insect larvae, particularly those drifting downstream in the water column. Everyone likes Brook Trout, but few like suckers. It is almost as if feeding on drifting insects versus feeding on bottom-dwelling insects is equivalent to eating on a table versus eating off the floor.

As a student of suckers, the late Linden Forest Edwards (1899–1970), a central Appalachian native of Lewisville, Ohio, and professor of anatomy and dentistry, succinctly summarized the mouth morphology of suckers: "The family Catostomidae (suckers) is among the best known and most easily recognized of the fresh-water fishes. They are characterized by the presence of a protractile mouth, fleshy, suckerlike lips, and toothless jaws, the mouth being protrusible to a greater extent than in other fishes."[10] Perhaps there is some circularity in characterizing a sucker as having suckerlike lips, but the lips of a sucker are somewhat unique. You were probably not planning to read about fleshy fish lips, at least not today. But here you are, about to be pondering the importance of sucker lips and their plicate or papillose structure. Plicate lips have parallel grooves, whereas papillose lips have a pimpled texture. In some sucker lips, both plicate and papillose textures are present (figure 10.2).

The plicate and papillose structures increase the surface area of the lip surface. A microscopic view of the lips shows that taste buds are blanketed across the ridge and valleys of the plicate lips, or across the peaks and valleys of papillose lips.[11] The increased surface area of plicate and papillose lips allows for an increased number of taste buds, thus greatly increasing taste sensitivity, and presumably food selectivity.

For most suckers, the lips are protractile and protrusible, and often located ventrally on the head, sometimes referred to as an inferior mouth

FIGURE 10.2. Ventral view of plicate lips (*left,* Golden Redhorse Sucker), papillose lips (*middle,* Northern Hogsucker), and lips in plicate-papillose combination (*right,* Torrent Sucker). Adapted and redrawn from Jenkins and Burkhead (1994), courtesy of the American Fisheries Society.

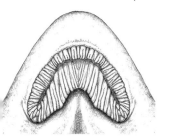 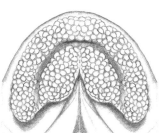 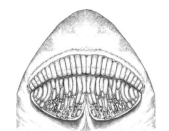

position. Bottom feeding in many species of suckers does not seem so unusual, particularly when considering this ventral mouth position. As Edwards explained in 1926, "In the act of feeding, the mouth is drawn downward and forward from the underside of the head like a spout. Food is taken into the mouth by suction, to which their protractile mouths and fleshy lips are peculiarly adapted."[12]

Not surprisingly, *Merriam-Webster* lists many meanings for "sucker," including (1) "any of numerous chiefly North American freshwater bony fishes (family Catostomidae) closely related to the carps but distinguished from them especially by the structure of the mouth which usually has thick soft lips," (2) "a person easily cheated or deceived," (3) "a person irresistibly attracted by something specified," and (4) "a lollipop."[13] "Sucker" is generally not a pleasantry when directed toward a person. Of those phrases that I can print, uncomplimentary ones include "sucker for punishment," "play (one) for a sucker," "there's a sucker born every minute," and "never give a sucker an even break." Further, no one likes to be sucker punched, sucker listed, or suckered in. Sucker synonyms are "fool," "dupe," "chump," "patsy," "sap," and "stooge."

Because of double meaning, or in some cases double entendre, it seems that suckers are doubly doomed from a public perception standpoint. In my many years of sampling fishes of central Appalachian streams, I have noticed some consistent patterns in public attitudes about suckers. For example, given the large amount of private land in the central Appalachians, I often have opportunities to talk with landowners about suckers, particularly after asking permission for stream access. On a few fish-sampling excursions each year, a common conversation unfolds as follows:

> Me: I'm writing a book on the fishes of the region, just wondering if you would allow us to sample fishes from the stream on your property.
> Landowner: All right with me, but just toss all of those no-good-for-nothing suckers up onto the stream bank.
> Me: [Expecting one of several possible answers, I ask the inevitable question:] Why?
> Landowner: Those suckers are not good because they suck up all the trout eggs / they muck up the water / they crowd out the trout / I can't catch them with rod and reel / I can't eat those bony things / I just don't like them.
> Then later, when my students and I begin sampling the stream, I get to answer the following question:
> Student: Are we going to do that?

Me: Do what?

Student: Kill all the suckers that we catch by tossing them onto the
stream bank.

Me: No.

Student: Why not?

Me: Because the landowner wants those *worthless* suckers
removed—but all the suckers that we catch are of tremendous
ecological value, particularly toward the health and function of
the stream.

Inevitably, two things follow. First, we catch a large Northern Hogsucker
(figure 10.3). Second, a young lady from the class holds this sucker upright
and plants a kiss right on its protruding lips. Every time this happens, I ask
the most obvious question: *Why does this happen?* I suspect that young biologists are excited about nature and fully appreciate all fishes, including those
ecologically important fleshy-lipped suckers. I have never seen a nonbiologist
kiss a sucker, and I think that I know the reasons why. As previously outlined
by the landowner, there is a long list of myths about suckers, including that (1)
suckers negatively impact other fish populations by eating eggs, (2) suckers are
somehow bad because of their bottom-foraging habits, (3) suckers outcompete
other fishes for food, (4) suckers are not a fish that you can catch with hook
and line, and (5) the bony bodies of suckers make them unpalatable.

John R. Greeley, an early twentieth-century ichthyologist and trout expert, was one of the first to dispel the egg-predation myth. He opened his
1932 paper on "The Spawning Habits of Brook, Brown, and Rainbow Trout,
and the Problem of Egg Predators" with these lines: "The problem of the
success of natural reproduction of trout is one of considerable interest to
the fisheries investigator. The question of destruction of trout eggs by various enemies, conspicuously the sucker and other so-called 'coarse fishes,' has
often been discussed before this Society."[14]

FIGURE 10.3. An irresistible Northern Hogsucker from a Central Appalachian stream.

Greeley summarized his findings as follows:

> The relation of the common sucker and other possible predators of trout eggs to trout reproduction was studied by field observations. . . . Female trout construct a pit and deposit eggs at the bottom of this, among coarse gravel or even large stones, during a single act of spawning. The nesting process is repeated several times before all of the eggs contained by one female have been deposited. Immediately after spawning, female trout cover the eggs with a thick coating of gravel. Defense by the male is continued during the early stages of nest covering while the female defends the redd for several hours after spawning.[15]

Greeley further outlined the methods necessary for suckers, sculpins (i.e., muddlers), or other fishes to eat trout eggs: "Any eaters of trout spawn must get the eggs by one or more of the following means: (1) by rushing in and securing eggs at the moment of deposition; (2) by digging out eggs after they have been covered; (3) by taking stray eggs which are not within a redd pit."[16]

Finally, Greeley emphasized that suckers, sculpins, and other fishes are unlikely to find and eat eggs, particularly those eggs within the interstitial spaces between pieces of gravel in the nest (redd). Eggs displaced outside of the redd are sometimes consumed, but these "stray eggs" or "waste eggs" would have died naturally by other means. Greeley wrote:

> Attempts of trout and muddlers to take eggs from the pit were success-fully prevented by female trout in the majority of observed instances. At most, a very slight percentage of the eggs deposited in the pit are taken in the interval between spawning and covering of the eggs. No attempts to dig out and feed upon eggs in the finished, covered redds were seen. By the time the female trout desert the eggs these are so well covered by gravel that disturbance by predators is unlikely. Waste eggs are common, due to the fact that some eggs fail to lodge in the pit and, because female trout often dig redds at areas previously used by other trout. The percentage of eggs which are loose in the stream rather than firmly lodged in covered redds is not large. The numbers are sufficient, however, to be sought by egg-eating fishes, notably the muddler, common sucker, brook trout, brown trout, and rainbow trout. Since eggs which are loose in the stream are unprotected from light and mechanical injury they are to be regarded as waste eggs, the destruction of which cannot be harmful to trout reproduction.[17]

Although dietary overlap does occur between some species of suckers and other fishes,[18] the bottom-feeding behavior of suckers often reduces competition with many other insectivorous fishes. As noted by Jenkins and Burkhead, "Their bottom-feeding specializations result in an inability to feed on drifting organisms that are important to so many other fishes; most suckers must feed on clean bottoms."[19] In one study, similar food items found in the diets of Black Redhorse Suckers and Smallmouth Bass were initially considered as evidence for competition, but then the researchers realized that the Smallmouth were likely eating benthic insect larvae dislodged by the bottom-foraging suckers.[20] This type of foraging association is not uncommon, and is sometimes referred to as "nuclear-follower feeding associations."[21] While river snorkeling, I have witnessed this nuclear-follower relationship many times, where a Northern Hogsucker dislodges larval insects while bottom foraging, and other fishes position downstream to feed on the resulting drift. This is known as a commensal relationship, because the foraging followers are benefiting from the actions of the sucker, but the sucker seemingly does not benefit from the followers.[22]

It is not true that suckers are worthless because you can't catch them by fishing. Suckers are catchable with hook and line, often requiring a high level of skill and a precise presentation, or in some cases you just need a bit of pure luck (figure 10.4).

FIGURE 10.4. A wonderful, wily White Sucker angled by the author. Photograph by A. Tomkowski.

It is also not true that suckers are too bony to eat. People have eaten suckers for a long time. Archaeological findings show that suckers were taken as food by First Americans at the end of the Pleistocene, around 13,500 years ago.[23] These Paleoindians perhaps had a paleo diet,[24] where spear points may have been used on fish instead of big game.[25] More recently, suckers have been bought and sold at fish markets throughout the greater Mississippi drainage. Three species of the genus *Ictiobus* (Bigmouth Buffalo, Black Buffalo, and Smallmouth Buffalo) have long been harvested commercially, often under different names. Bigmouth Buffalo are often referred to as "Gourdheads," whereas Black Buffalo are caught and sold as "Blue Rooters," and Smallmouth Buffalo are known as "Razorbacks." Interestingly, none of these common names seem particularly appetizing. Perhaps a name change would be effective at increasing sales, an approach that worked well for the Slimehead, a marine fish whose name was changed to Orange Roughy.

After Buffalo are bought at local markets, the focus is often on cooking and eating the ribs, a meaty section of the fish without a lot of small bones. In other cases, bones of suckers can be minced or removed. For example, efforts to market the meat of White Suckers have included canned minced sucker, smoked sucker sausages, mechanically deboned sucker, vacuum-packed sucker, and, in an extreme case, sucker hotdogs.[26]

My preferred method of cooking is deep frying, but you must fleece the sucker first. Although "fleecing a sucker" sounds like someone getting ripped off by a scam artist, it is actually a method of scaling a fish (i.e., removing the scales). Use a sharp knife in a slow-sawing motion to slice off the scales on the side of the dead fish. Start at the tail and work toward the head. The end result is a scaleless sucker with skin still attached. Fillet the fish after fleecing, but do not remove the skin. Next, score the fillet by placing the skin side down on a cutting board and make a series of downward cuts, down to the skin, but without cutting the skin. Next, I like to cut the fleeced and scored fillet into two-by-two-inch (five-by-five-centimeter) pieces. Beat an egg and dip each piece of scored fillet in the egg batter. Then cover the egg-batter-coated pieces of fillet with cornmeal, making sure that the cornmeal gets into the scores of the meat. Finally, submerge each piece in hot peanut oil (375°F or 190.5°C). The bones are cut into smaller segments by the scoring process, and are subsequently eliminated by the hot oil.

Myths and misconceptions about suckers have swayed public opinion, misaligning these species in the minds of many. In an attempt to set the record straight, this story emphasizes several reasons as to why we should value these bottom fishes. First, as the case with "bottom-feeder," the word "sucker" was a biological term for fishes long before its negative applications of usage toward people. Don't be swayed by the name.

Second, although suckers may eat a few fish eggs, they do not generally have negative effects on the reproductive success of other fishes. If suckers were bad for stream ecosystems, then you would expect that streams swimming with suckers would have few other species. As noted for Virginia streams by Jenkins and Burkhead: "Suckers do not appear to depress community diversity. Rather, they are integral parts of communities; the most diverse fish faunas of Virginia include 5–11 species of suckers."[27] It would be accurate to reword the previous sentence as "The most diverse fish faunas of streams of the central Appalachians include many species of suckers." Suckers benefit aquatic communities. They play important roles as forage fish to aquatic and terrestrial piscivores, as bottom foragers that free up food for drift feeders, and as subsidizers of nutrients in upstream river reaches.

Third, eating suckers or fishing for suckers in central Appalachia is currently not common, but I predict an increase in future interest. You are not likely to find sucker fillets in the seafood section of the grocery store, at least not any time soon, and neither will you commonly find fishing lures designed and marketed to catch suckers. But consider catching and eating a sucker from a clean stream—you may be surprised by the outcome. Or better yet, savor the satisfying culinary experience of cooking a sucker over a campfire—a "sucker on a stick" scenario that occurred in the central Appalachians long before the lollipop.

Madtom Miniatures

Common sayings are often contradictory. Consider "Good things come in small packages" versus "The bigger the better." We often place values on relative terms, such as small and big, although these can differ among us. Once I asked two of my daughters, ages four and five, to arrange four paper fish shapes of different sizes. The older daughter placed the two larger fish above the two smaller ones. Her explanation: it was two parents caring for their two offspring. My younger daughter surprised me with her arrangement—a row of fishes ordered from large to small. Her explanation: big fish eat little fish.

In nature, some fishes benefit from being small. The tiny Tippecanoe Darter is a testimony to small-sized success (see chapter 22, "Tippecanoe Is Tiny Too"). The Tippecanoe's small size allows it to live in small spaces of gravel beds in shallow riffles. The benefits of smaller size, however, may be lost in the presence of bigger predators. As my young daughter surmised, small fish are eaten by big fish. The risk of predation is generally high for juvenile and small fish. As a contemplator of predator-prey relationships, I prefer not to place a value on prey and predators in terms of good or bad. My thoughts do, however, often go to a single question when pondering small fish—how do small fish avoid predation?

In biological terms, predators of fish are properly labeled as "piscivores," a category covering a wide range of amphibians, fishes, reptiles, birds, and mammals. In the central Appalachians, the American Bullfrog and the Eastern Hellbender (a large aquatic salamander) are two amphibians that commonly feed on small fishes. The Common Watersnake, Queen Snake, and Snapping Turtle represent three aquatic piscivorous reptiles. Several birds eat fishes, including the Green Heron, Great Blue Heron, Belted Kingfisher, Bald Eagle, and Osprey. There are also several fish-eating mammals—Fisher, River Otter, and Raccoon.

Some piscivores, particularly fishes, are mouth-gape limited, where the size of their mouth limits the size of their prey. Take a moment and imagine yourself as a fish predator. You could easily swallow a small fish up to 2 inches (5.1 cm) in length, like an anchovy, a sardine, or a goldfish. Swallowing a larger fish whole, like a 12-inch (30.5-cm) Brook Trout, would not work. But your ability to bite off and swallow smaller pieces allows you to eat bigger fish. Some piscivorous birds also eat big fish. For example, a Bald Eagle or Osprey can use its stout beak to tear off small strips of flesh from a large fish. However, other fish-eating birds, like herons and kingfishers, are gape-limited and must swallow a whole fish.

Most fishes are small during adolescence and thus vulnerable to larger predators. Many fishes reduce predation risks by growing bigger—small juveniles grow into a larger adult size. Some fishes do not attain a large adult size and are at high risk of predation for their entire life. As an example, consider the madtoms, sometimes called catminnows, willow cats, douglers, and beetle-eye, a group of catfishes of small adult size. On average, most adult madtoms are under 3 inches (7.6 cm) in length. Madtoms are often used by anglers as fish bait. Over 25 species are native to North America. The central Appalachians host at least 6 species, including the Brindled Madtom,

FIGURE 11.1. A 1.7-inch (43-mm) Brindled Madtom (*top*) and a 4.8-inch (122-mm) Margined Madtom (*bottom*)—representatives of the many miniature madtoms of the Central Appalachians.

Margined Madtom, Mountain Madtom, Northern Madtom, Tadpole Madtom, and Stonecat Madtom.

Madtoms have the general appearance of most catfishes (figure 11.1)—"whiskers" around the mouth, smooth skin without scales. A common misconception is that catfish whiskers are poisonous stingers. The whiskers, also known as barbels, actually assist in finding and tasting food. The madtom's body, like that of most catfishes, is widest at the back of its head, near where the skull connects with a well-developed and rigid skeletal frame known as the pectoral girdle.

Like many fishes, catfishes are flanked with fins. Fish fins are often supported by a series of rays or spines. In madtoms, the dorsal fin begins with a short platelike spine followed by a long hard spine and a series of soft rays. The second dorsal fin, known as an adipose fin, is ray-less and fleshy. The anal fin, caudal fin, and pelvic fins are made up of soft rays without spines. The pectoral fins are large relative to the body size, and each pectoral fin begins with a long, hard, sharply pointed spine. Catfish spines are covered by a sheath of skin. Some species have a serrated edge on the backside of the pectoral spine—a teethlike row of bony points that can be straight or recurved (figure 11.2).[1]

Madtoms reduce their risk of predation in several ways, many shared with other small-bodied fishes. One way a small fish avoids predation is to avoid being seen. Several methods of predator avoidance by Logperch are discussed in chapter 18, "Like a Rolling Stone," including countershading,

FIGURE 11.2. Pectoral fin spines of six species of madtoms from the Central Appalachians; (a) Tadpole Madtom, (b) Margined Madtom, (c) Stonecat Madtom, (d) Brindled Madtom, (e) Mountain Madtom, and (f) Northern Madtom. The base has three lobes as numbered on the Tadpole Madtom spine. Adapted and redrawn from Taylor (1969), courtesy of the Smithsonian Libraries and Archives, Smithsonian Institution, and NMNH Division of Fishes.

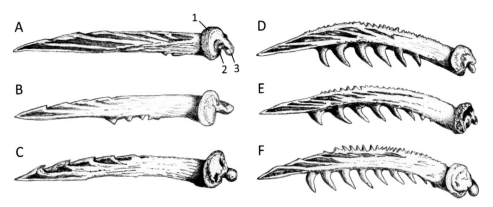

background matching, and disruptive coloration. Similarly, madtoms are often countershaded with dark backs and lighter undersides. Dark backs blend with the background of the stream bottom. Some are disruptively colored with dark dorsal saddles of pigment on their backs.

Madtoms are mostly nocturnal. During the day, madtoms find refuge on the stream bottom under leaves, logs, or rocks. Often madtoms take refuge under the largest available object, natural or not. Unfortunately, streams of the central Appalachians are often unnaturally littered by uncaring individuals—I have often found madtoms hiding under or inside of submerged truck tires. Once I even found one under a partially submerged couch.

Although madtoms reduce predation risks through camouflage and daytime refuge, some predator encounters are inevitable. Approached by a predator, a madtom sometimes does what many other small fish do—erect its fins to give the impression of a larger body size. Most piscivores know their limitations, and may pass on a prey that appears larger than their gape size. Fishes erect fins by extending them outward to about 90° with the body, known as fin abduction. Through abduction of the dorsal and anal fins, a fish appears to have a deeper body. The erection of pectoral fins along the side of the fish gives the appearance of a wider body. In madtoms, an extended dorsal fin approximately doubles body depth and extended pectoral fins approximately triple body width.

With fins erect, the diminutive madtoms, like most catfishes, have an added predator deterrent—the hard needle-sharp spines of the pectoral and dorsal fins. Most fishes erect fins when stressed by a predator's presence, but if captured, the predator can depress the prey's erected fins by swallowing the prey headfirst. In catfish, however, the pectoral and dorsal spines can be locked in place following full fin abduction, making it difficult for a predator to swallow.

In order to understand the mechanism of spine locking, we need detailed description of the base of the spines. At the base of the pectoral spine there are three lobes, the dorsal, anterior, and ventral processes (see figure 11.2).[2] The pectoral girdle, a well-developed and rigid skeletal frame located just behind the head, is also a critical part for spine locking. When the pectoral spine is erected, the three processes at the spine base articulate (form a joint) with the pectoral girdle. During spine locking, the three processes at the spine base bind and lock with the pelvic girdle through friction and an interlocking joint.[3]

Interestingly, the dorsal and pectoral spines of most madtoms are surrounded by venom glands, which occur near the spine base or surround the spine shaft. Early on, some researchers incorrectly assumed that the venom was forced out the end of a hollow spine shaft, like a hypodermic needle.[4] Instead, the venom flows through grooves in the surface of the spine shaft. Muscles do not squeeze out the venom, but the venom flows freely after the sheath of skin is removed from the spine tip. When a spine pierces a predator, the skin sheath surrounding the spine and venom glands is severed and venom enters the wound.[5]

A common tendency is to treat prey as underdogs and predators as villains. After a closer look at the madtom's arsenal of antipredator defenses, you may become less concerned about its safety and security. In fact, your concern may become centered on the safety of the predator! Seemingly, it should be difficult for a predator to swallow and stomach a madtom. However, madtoms and other catfishes are common diet items for many piscivores.

It seems reasonable that predators should completely avoid or carefully handle dangerous prey, such as madtoms.[6] As a consequence, dangerous prey should make up a relatively lower percentage of a predator's diet. Within the central Appalachians, Great Blue Herons are a common wading bird known to eat madtoms and other catfishes, although few studies have focused on catfish as prey for birds of the central Appalachians. A diet study of Great Blue Herons at two sites in southeastern British Columbia found that Black Bullhead catfish made up 51% and 25% of prey items.[7] At the same sites, Black Bullhead catfish made up 82% and 50% of prey items in Osprey diets. Ospreys, however, are not gape-limited, and tear off small pieces of fish; hence increased handling associated with spines is not an issue.

Common Watersnakes, a common aquatic snake of central Appalachian streams, also eat madtoms. Although watersnakes are somehow effective at swallowing madtoms, the snake does not always fare well in the digestion process—several references report catfish spines protruding out of the sides of watersnakes.

It seems contradictory that madtoms are commonly used as fish bait, but are also well equipped with antipredatory weapons of spines and venom.[8] In an experiment, Largemouth Bass were injected with venom of Tadpole Madtoms.[9] The effects on the bass included color loss, muscle twitching, hemorrhage, and loss of equilibrium. Symptoms occurred immediately but subsided within an hour. Another study found that the presence of spines increased catfish survival, but did not reduce vulnerability of attack by Largemouth Bass.[10] This study also reported an increase in prey-handling time—the amount of time between attack and ingestion. Largemouth Bass typically attack and disgorge catfish multiple times before finally finishing. Another study found that Largemouth Bass had little trouble swallowing madtoms head first, but had handling times of up to two hours when attempting to swallow one tail first.[11] From an angler's perspective, catfish are excellent bait because of their ability to survive multiple attacks from predators—they stay alive for a long time while on the hook. Fortunately for anglers, Largemouth Bass are often patient with the longer handling times required for dangerous prey, such as madtoms.

Not surprisingly, anglers often suffer from being "spined" on the hand while putting a madtom on a fish hook. The pain from this has been likened to the stings of bees and hornets. One treatment option is to wash the wound with hydrogen peroxide.[12] Another quick treatment is immersion of the hand in extremely hot water, or a hot water soak with Epsom salts,[13] as the heat reduces the pain.[14] Some have suggested that urinating on the wound will relieve the pain, although I have not tried this treatment. In medical journals, there are quite a few scientific articles on catfish wounds.[15] Madtom stings are not extremely dangerous—symptoms typically include severe pain and swelling at the site of the sting. Secondary infection, however, is a concern, particularly if part of the spine stays in the wound.

Catfishes, although not the only venomous fishes, may make up the majority.[16] Catfishes occur on every continent except Antarctica, and include about 1,250 venomous species.[17] Other venomous fishes, approximately 850 species, include representative species of sharks, rays, toadfishes, stargazers, weeverfishes, blennies, jacks, rabbitfishes, surgeonfishes, scats, gurnard perches, scorpionfishes, and stonefishes.[18] Although the Saber-Toothed Blenny (Indian and Pacific Oceans) has venomous canine teeth, most venomous fishes have venom glands associated with fin spines.

In addition to treating wounds, medicinal researchers are also interested in venom from a pharmacological perspective.[19] Though toxic, venoms often contain compounds that have medicinal uses. Examples include compounds of the venom of bees and cobras for treatment of arthritis, the venom of scorpions used for cancer treatment, and tarantula venom for the study of muscular dystrophy. I predict that compounds in fish venoms will also have medical value. The potential for medicinal uses of animals is one of many reasons to conserve biodiversity—a loss of a species to extinction could mean the loss of potential future medicinal discoveries.

I hope that my chapter on madtoms will lead you to value small fishes of the central Appalachians. Our little madtoms are amazing in their anti-predator advances—particularly their armature of spines and venom. At this juncture, I will not suggest that you urinate on a madtom wound. Also, I do not suggest jamming a madtom spine into one of your arthritic joints, but I will not be surprised if future research celebrates the "medicinal madtom." Perhaps while waiting for advances of science, you could read more about madtoms, conduct your own studies by keeping madtoms as aquarium pets (if your government regulations allow), or just sit back and muse about the misconception of small-sized animals being the underdogs of nature.

Trees and Trout

You don't have to be a naturalist to be enamored with nature. Consider Frank Lloyd Wright, a famous architect who nurtured a keen interest in nature. Wright incorporated nature in his organic architectural designs. He often embodied an ecosystem theme in his work, in which nonliving components of the environment were intricately linked with a community of living organisms. A prime example of Wright's work is the Fallingwater House, a summer home built during 1936–37 on top of a waterfall on Bear Run of the Laurel Highlands of central Pennsylvania.[1] A National Historic Landmark, it is perhaps the most celebrated architectural design of the central Appalachians.

With its placement on top of a waterfall, the Fallingwater House syncs with the stream (figure 12.1). Both house and stream are nested within a riparian area of Eastern Hemlocks and Rhododendron. The word "riparian" is from the Latin word *ripa,* meaning riverbank, referring to the interface between stream and land. Within Fallingwater House, the cool running water of the stream provides a natural air-conditioning, while broadcasting the "song of the stream," which in my mind is one of the most beautiful sounds of nature. From an exterior view, the house has a series of cantilevered decks, which flow naturally with the waterfall and the stream, and harmonize with the horizontal branches of the riparian Hemlock trees.

Bear Run, the stream that flows underneath Fallingwater House, supports a population of native Brook Trout (Figure 12.2). Millions of people have visited Fallingwater House, but most were completely unaware of Brook Trout in the stream. Yet the native Brook Trout is perhaps the most celebrated fish of the central Appalachians, serving as the state fish of Pennsylvania, Virginia, and West Virginia. Also, the Eastern Hemlock

FIGURE 12.1. Photograph of Fallingwater House, courtesy of the Western Pennsylvania Conservancy.

holds the honor of being state tree of Pennsylvania. From a naturalist's perspective, the Brook Trout and Fallingwater House share a commonality in that both are interconnected with the stream, the Hemlocks, and the riparian landscape.

Whether you are touring a historic landmark, hiking, or fly fishing for native Brook Trout, a visit to a central Appalachian mountain stream is an unforgettable experience. People experience nature in different ways, but most interactions with nature are the sum of the senses: sight, hearing, touch, and smell. The centerpiece of the mountain stream experience is the loud but calming, gurgling song of the stream. Along the stream banks, moist moss-covered boulders and tree roots glisten intermittently from an interplay of shadows and narrow beams of light, kaleidoscoped by the canopy cover of the forest trees. Eastern Hemlocks and several species of deciduous hardwood trees tower over an understory of young Hemlock trees and Rhododendrons. The Hemlocks, Rhododendron, and cold water all contribute to the cool, fresh scents of the mountain air.

FIGURE 12.2. Photograph of a native Brook Trout. This fish was about five inches (12.7 cm) in length.

Unfortunately, many of our Eastern Hemlock trees are dying, owing to an introduced insect, the Hemlock Woolly Adelgid.[2] This small aphid-like insect, which is native to Japan, is reddish brown to purplish black when young, but becomes woolly white as it matures, reaching a length of about 0.06 inches (1.4 mm).[3] Basically, these little buggers are sap suckers that deplete starch reserves, thus weakening or killing the Hemlock trees. A large loss of Hemlock trees is a major concern on many levels. How will the loss of Hemlock trees influence both the nature experience and the Brook Trout populations of central Appalachian mountain streams?

The sight and scents of large Hemlocks are a huge part of the mountain stream experience. For me, most memorable are the majestic older Hemlocks, tall and evergreen, with broad, dense canopies. The Hemlock is often referred to as a foundation species, because it influences biodiversity and ecosystem function.[4] Riparian Hemlocks are responsible for creating a unique cool microclimate of the mountain stream ecosystem.[5] The Hemlock canopy shades and insulates the stream, keeping air temperatures cool during summer and warmer than the surrounding forest during winter. The shade produced by the large evergreen canopies of the Hemlocks also reduces water evaporation from the stream and riparian area. Trees transpire water, a process where moisture moves upward inside of a tree, evaporating from stems and leaves. During the dry period of summer, riparian Hemlocks transpire less water than deciduous hardwood trees.[6] During the warm summer season, evapotranspiration (the combination of evaporation and transpiration) in an area of riparian Hemlocks is less than that of hardwoods.

Population persistence of Brook Trout is often determined, in part, by summer conditions, a time when levels of stream flow can get low and water temperatures can exceed survival thresholds. In the central Appalachians, native Brook Trout persist in cold-water streams, and generally are not found in streams where water temperatures exceed 68°F (20°C),[7] preferring temperatures near the range of 57.2°F–60.8°F (14°C–16°C).[8] Often Brook Trout streams in the central Appalachians are at higher elevations or are fed by groundwater springs. Usually stream flows are relatively high during spring and winter, and low flows in fall are often offset by cooler temperatures. During droughts, pool habitats serve as a refuge for Brook Trout, particularly if the pool is shaded by canopy cover. High-gradient streams have plunge pools at the base of cascades or waterfalls, whereas pools in lesser gradient streams are often associated with submerged or partially submerged dead wood. The dead wood in the stream provides important habitat for Brook Trout in many central Appalachian streams.[9] Despite pool presence, severe droughts can completely dry the stream, resulting in extirpation of Brook Trout within that stream.

Groundwater sources and rainfall contribute to stream-water flows, but trees also have an influence. The loss of riparian Hemlocks and associated changes to evapotranspiration rates will lead to seasonal changes in stream flows.[10] The loss of Hemlock trees along the stream will create canopy gaps, allowing increased levels of sunlight to reach the water. A study near Harrisonburg, Virginia (Fridley Run, George Washington National Forest), found that the loss of riparian Hemlocks increased light levels by as much as 65%, but had little influence on stream-water temperatures owing to the cooling effect of the groundwater input.[11] In the mountains, topography and aspect differ among watersheds and influence the amount of sunlight reaching a stream, thus streams will be influenced differently by the loss of Hemlocks.

Hemlock loss may also alter stream-water chemistry. An increase in light and soil temperature is expected to increase rates of both decomposition and nitrogen cycling, thus influencing nutrients in the stream.[12] As nitrates are leached into the stream, water concentrations of calcium and aluminum are expected to increase.[13] Many streams in the central Appalachians could benefit from a pulse of calcium, but Brook Trout populations are generally greatly reduced when aluminum concentrations are elevated.[14]

The loss of Hemlock trees will influence Brook Trout diets. Brook Trout are primarily insectivores, feeding on aquatic insect larvae that live in the stream as well as larvae and adults of terrestrial insects that fall into the stream.[15] In addition to insects, Brook Trout also eat other arthropods, such as spiders. Several research studies have demonstrated that streams draining Hemlock forests often have lower densities of aquatic insects than those of hardwood forests.[16] The counterside is that Hemlock trees harbor a diversity of arthropods,[17] providing an important terrestrial food source for Brook Trout. One study found that Brook Trout were three times more abundant in watersheds with Hemlock trees than in those dominated by hardwoods.[18]

Several studies from across Appalachia have reported from 250 to 550 species of arthropods associated with Hemlock trees,[19] a finding that supports a diversity of food items available for potentially falling out of lower branches and into the water. Sweka and Hartman reported that terrestrial invertebrates may exceed 50% of Brook Trout energy consumption.[20] Utz et al. found that terrestrial beetles comprised nearly 40% of Brook Trout energy consumption during the months of May and June.[21] One study found that the diversity of moth species was higher in Hemlock forests than that in neighboring non-Hemlock forests,[22] suggesting that Brook Trout in Hemlock forests may have access to a higher diversity of caterpillars (i.e., moth larvae) and adult moths, which may fall into or fly into the water, respectively. The take-home message is that a lot of arthropods live on riparian Hemlock trees, fall onto the water surface, and become a meal for Brook Trout.

Hemlocks and Brook Trout have coexisted for a long time, demonstrating a resilience to environmental changes. Both species have experienced periods of climate warming and cooling and periods of wet and dry conditions. First, consider a timeline of Hemlocks during the late Pleistocene and Holocene. The Pleistocene epoch, often referred to as the Ice Age, was a period of time that lasted from approximately 2.58 million years ago until about 11,700 years ago. The Holocene is the period of time from about 11,700 years ago to the present. Based on fossil pollen from eastern North America, Hemlock trees represented less than 1% of all trees from approximately 21,000 to 15,000 years ago, but exceeded 10% in some areas of the central and southern Appalachians at about 14,000 years ago.[23] The distribution of Hemlock trees expanded northward, where the species reached its highest densities around 6,000 years ago (figure 12.3).[24] At approximately 5,000 years ago, Hemlocks suffered a drastic die-off, where populations did not recover in some areas for nearly two millennia (figure 12.3).[25] Several hypotheses have been considered for the mid-Holocene Hemlock die-off, including severe drought conditions,[26] an infestation of the Hemlock Looper moth,[27] and a forest pathogen.[28] Since then, populations in the central Appalachians have never recovered to previous densities. Given the current threat of Hemlock Wooly Adelgid, it looks as if Hemlock populations could reach a new low.[29]

At present, I can only speculate on Pleistocene and precolonial Holocene changes to the distribution of Brook Trout in the central Appalachians. Most likely, distributions expanded during periods of cooling and contracted during warming. Climate reconstruction has predicted a sinuous cycle of cold to warm periods during parts of the earth's history.[30] As

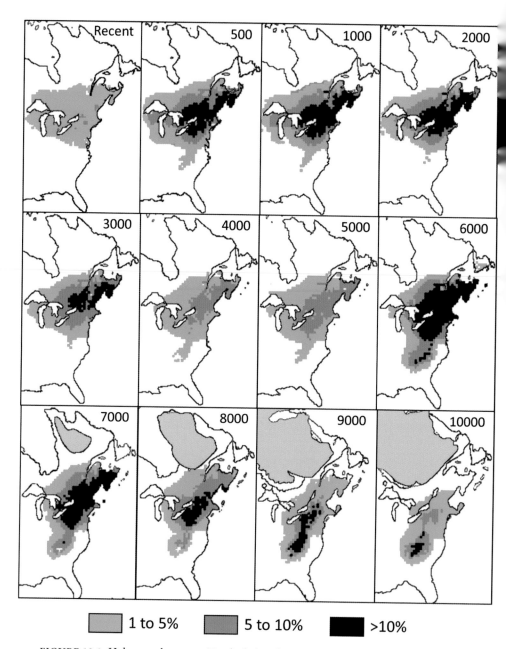

FIGURE 12.3. Holocene changes to Hemlock distributions in eastern North America inferred from percentage fossil pollen. Panels range in time from recent to about 10,000 years ago. Blue areas represent ice. Modified from Williams et al. 2004, https:// esajournals.onlinelibrary.wiley.com/doi/epdf/10.1890/02-4045; http://creativecommons .org/licenses/by/4.0/.

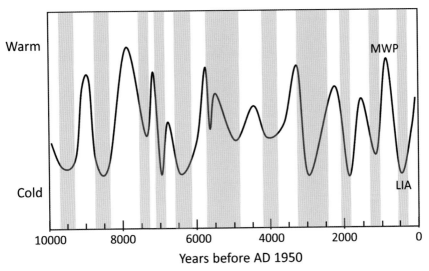

FIGURE 12.4. Speculations on warm and cold periods during the Holocene from about 10,000 years ago to AD 1950 based on fossil pollen (using an approach similar to that of Viau et al. 2006), with colder periods (gray bands) associated with glacial advance, climate cooling, and rapid climate change events. Two recent events are highlighted: the Medieval Warming Period (MWP) and the Little Ice Age (LIA).

an example, consider the cycles of expansion and contraction of glaciers during the Pleistocene. The Holocene period, however, has also experienced cold and warm periods, in part owing to glacial readvancement, cycles of solar activity, and variations of the earth's orbit (figure 12.4).[31] I can speculate about warm and cold periods of the Holocene based on fossil pollen (using an approach similar to that of Vaiu et al. 2006),[32] with colder periods (gray bands in figure 12.4) associated with glacial advance,[33] climate cooling,[34] and rapid climate change events.[35] Currently we are in a period of warming following the Little Ice Age, a cold period lasting from about 530 years ago to the early 1900s. More recently, there is a concern about current and projected levels of greenhouse gases in the atmosphere, and the predicted potential for future warming.[36] Current temperatures are near those of the Medieval Warm Period, a time from approximately 1,120 to 530 years ago occurring just before the Little Ice Age. If the climate of the central Appalachians continues to warm, then Brook Trout populations will likely continue to contract.

Some things can only bend so far before they break. After surviving the cycle of climate swings of the Pleistocene and precolonial Holocene, the eastern Brook Trout has been bullied by environmental degradation for the past 200 years. Brook Trout have been extirpated from many watersheds of the central Appalachians and other parts of the range, resulting in today's fragmented distribution. Many contributing factors have likely led to reduced or extirpated populations. From about 1880 to 1920, logging companies cleared the virgin timber of the central Appalachians, including the Hemlock trees.[37] Logs were often moved downstream within stream channels by using splash dams. Shay locomotives were later used to move logs, and were a main source of fires that burned many areas within the central Appalachians.[38] More recent impacts on Brook Trout include increased hydrogen ion concentrations (i.e., low pH) associated with acid mine drainage and acid rain,[39] competition with nonnative fishes,[40] urbanization,[41] dams and hanging culverts,[42] stream sedimentation,[43] and processes associated with shale gas extraction by hydraulic fracturing.[44] These recent impacts often occur at local or regional scales, but one study looked at cumulative impacts over the range of Brook Trout within the eastern United States (figure 12.5). Brook Trout were more likely to be extirpated from subwatersheds where human use of land exceeded 18%. Further, subwatersheds with less than 10% human use were more likely to have more than 50% of habitat intact and still support self-sustaining populations of Brook Trout.[45] It appears that our land-use practices are not very compatible with the conservation of Brook Trout.

The loss of Hemlock trees from the riparian area will drastically alter my nature experience, although the more important effect will be the alteration of ecosystem structure, including the impact on Brook Trout populations. Considering population losses of Hemlock Trees and Brook Trout, I find consolation in the following words from Frank Lloyd Wright: "Study nature, love nature, stay close to nature. It will never fail you." However, it becomes increasingly difficult to stay close to nature when important parts of your expected nature experience begin to disappear, such as Hemlock Trees and Brook Trout. Nature is connected in noncasual ways, as the sights, sounds, and smells of a central Appalachian mountain stream, which are integral to the nature experience, are also directly linked to critical habitat components of the native Brook Trout. My future visits to Fallingwater House will just not be the same if the Hemlock trees die and Brook Trout are extirpated from the stream. However, I do not think that the situation will go to that extreme. Populations of Brook Trout and Hemlock Trees will likely be reduced further during our current warming trend, but I predict that the Brook Trout and Hemlock Trees will win out in the end, persisting into the next Ice Age, as both have proven to be survivors over the long haul.

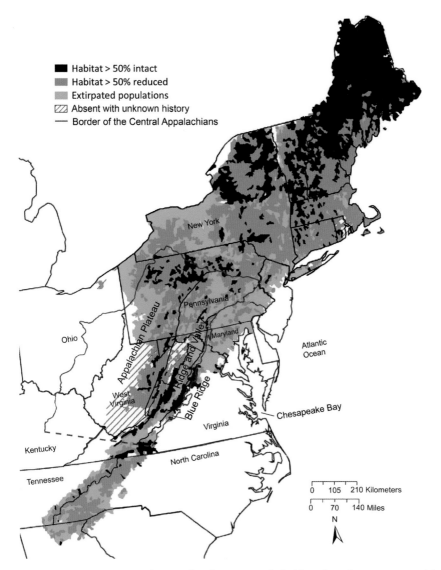

FIGURE 12.5. Eastern Brook Trout distribution map (USA) based on three categories: (1) black areas—historic habitat is intact and > 50% supports or is predicted to support self-sustaining populations; (2) dark green areas—historic habitat has been reduced and > 50% does not support or is predicted to not support self-sustaining populations, and (3) light green areas—watersheds where Brook Trout have been extirpated (i.e., no longer exist) or are predicted to have been extirpated. In some areas with species absence, extirpation is a possible explanation, but historical occurrence records are lacking. The red polygon represents the border of the Central Appalachians as defined in the preface. Brook Trout data are modified from Hudy et al. 2008, courtesy of the American Fisheries Society.

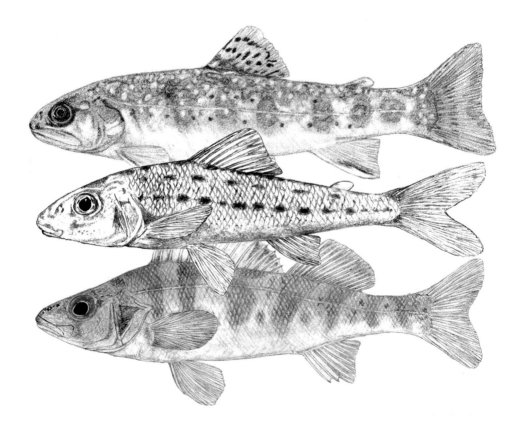

Appalachian Apposition

Many naturalists are compulsive collectors. Animal and plant specimens, as well as other items of scientific interest, are often squirreled away, amassed in a single space. This practice dates to at least the sixteenth century, when rooms of biological bounties were called cabinets of curiosities. Consider the cabinet of Ole Worm (1588–1654), a seventeenth-century physician and natural historian from Renaissance Scandinavia, who also went by the latinized name of Olaus Wormius (figure 13.1).[1] Cabinets of curiosities contained many interesting and unique conversational pieces, including exotic and mythical creatures. Another well-documented cabinet was that of Albertus Seba, an eighteenth-century Dutch zoologist, which influenced the scientific contributions of Peter Artedi (Petrus Martini Arctaedius) and Carl von Linné (Carolus Linnaeus).

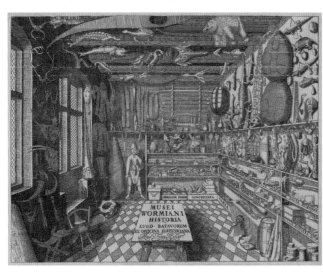

FIGURE 13.1. An engraving of Ole Worm's cabinet of curiosities (Musei Wormiani Historia) from the frontispiece of the *Museum Wormianum* (Worm 1655).

115

Artedi, a Swedish naturalist known as the father of ichthyology, was the first to introduce the concepts of genus and species with his work on classification of fishes. Linnaeus, a Swedish physician and naturalist known as the father of modern biological taxonomy, modified Artedi's concepts to develop the binomial Latin name format that we use today. The first word in a Latin binomial is the genus name, followed by the specific epithet. The binomial Latin name is the gold standard in published species descriptions, which are peer-reviewed publications that include, in part, the morphological characters of a species that are useful for distinguishing it from other species. The scientist credited with a description of a new species is considered as the authority on the species, and his or her name is often printed following the Latin binomial. As example, consider yourself as a member of the species *Homo sapiens* L., where "L." is an abbreviation for Linnaeus, who described our species in 1758.[2]

Cabinets of curiosities have often contained specimens of scientific interest. Some of these specimens have contributed to descriptions of new species, whereas others are unique or unusual in some way. According to my wife and other critics of "well-organized" office spaces, I am apparently the proud owner of two cabinets of curiosities, one at home and the other at my workplace. Both office spaces have some similarities: desks and floors covered with teetering stacks of journal articles, walls shelved with books, and field equipment from previous and current projects. Specimens are scattered about the rooms, including jars of various fishes, crayfishes, and aquatic insects.

Just recently, I noticed an interesting fish specimen on my home office shelf, one that I had collected from a small creek of the Little Kanawha River drainage, a tributary of the Ohio River. The specimen was submerged inside a small jar of ethanol, labeled with the Latin binomial *Percopsis omiscomaycus* (figure 13.2). It has an adipose fin, a small fleshy fin that occurs between the dorsal and caudal fins. This species, with individuals usually not exceeding four inches (10.2 cm) in length, is distributed throughout much of the northern United States and a wide part of Canada, but has part of its southern distribution in some streams of the central Appalachians.

Percopsis omiscomaycus has an interesting taxonomic history, and a slightly confusing common name—Trout-perch. It sounds like an oxymoron—is it a trout or a perch? Trout and perch represent two distinct groups of fishes. Species of trout are similar in many ways to those of salmon, and both are members of the fish family known as Salmonidae. Trout have cycloid scales, an adipose fin, and a single soft dorsal fin (figure 13.3). The perches are placed in the family Percidae. They have ctenoid scales and two dorsal fins, a spiny fin and a soft fin (figure 13.3), and no adipose fin.

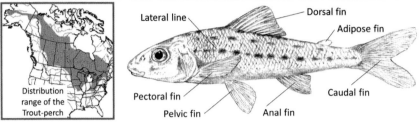

FIGURE 13.2. A photograph of a 2.9-inch (74-mm) Trout-perch, including the geographic distribution range and an illustration to emphasize the location of fins.

I start this taxonomic tale with Zadock Thompson, a naturalist, professor, and Episcopal minister from Vermont who once found an interesting dead fish on the shoreline of Lake Champlain in 1841. Thompson thought that this fish was new to science, and in his notes gave it the genus name *Salmoperca*. This name reflected his opinion that the fish contained characteristics of both trout (*Salmo*) and perch (*Perca*). He did not publish a species description at the time, because he had only a single specimen. Thompson stated that he later presented the specimen to Louis Agassiz, a Swiss American geologist and ichthyologist who was a leading expert at that time on fish classification (see chapter 14, "Fragmented Thoughts"). Thompson wrote, "When Prof. Agassiz was at Burlington, in 1847, I submitted the above mentioned specimen to his inspection, having at that time obtained no others. At first sight, he thought it might be a young fish of the salmon family, but,

FIGURE 13.3. A Brown Trout (*Salmo trutta*) and an illustration of a cycloid scale as representatives of the family Salmonidae (*top*); and a Yellow Perch (*Perca flavescens*) and ctenoid scale as representatives of the family Percidae (*bottom*).

upon further examination, he said it was not a salmon, nor any other fish with which he was acquainted."[3]

During summer 1847, Thompson obtained three more specimens of this fish. In September 1847, at the meeting of the Association of American Geologists and Naturalists in Boston, he presented one of the specimens to David Humphreys Storer, a well-respected physician and naturalist, in hopes that he would find out if anyone knew of this unusual fish.

Thompson, who subsequently obtained several more specimens of this interesting little fish, stated: "In May, 1849, I obtained from Winooski river a number of living specimens, which I kept alive for some time; and, observing the great translucency of the living fish, when held up towards the light, I gave it the specific name of *pellucida,* having previously called it, in my journal, *eoceta,* from its wing-like pectoral fins."[4]

On November 18, 1848, Agassiz gave a presentation on two new genera of fishes from Lake Superior at a meeting of the Boston Society of Natural History.[5] Twenty-four members of the Society were present, including the chair, David H. Storer. Agassiz placed perches in a group of fishes known as the percoids. The published meeting notes included the following:

> It has the adipose fin of the Salmonidae, but not the jaws of that family; these strongly resembling those of the Percoids. In its scales, which are serrated on their margins, it also resembles the Percoids. Its characters are sufficiently peculiar to justify the establishment of a new family from this single species.
>
> Prof. A. [Agassiz] exhibited a colored drawing of the new fish of which he was speaking, by Mr. J. E. Cabot; and presented specimens for the cabinet of the Society. He has given it the name "*Percopsis,*" on account of its resemblance to the Percoids.[6]

Subsequently, Zadock Thompson discovered that his genus *Salmoperca* was the one proposed as the genus *Percopsis* by Louis Agassiz. Thompson wrote:

> I noticed, in the proceedings of the Boston Society of Natural History, that Prof. Agassiz had laid before the Society an account of a new genus of fishes discovered by him in Lake Superior, which he proposed to call *Percopsis.* Suspecting, from the brief description given of it, that it was identical with my *Salmoperca,* I wrote to Dr. Storer and inquired of him, if the specimens from Lake Superior, presented to the Society by Prof Agassiz, were like the one I put into his hands in 1847. He wrote me that he could not say—that the specimen went out of his hands soon after he received it, and he had not seen it since.[7]

But the following year, from a July 18, 1849, meeting of the Boston Society of Natural History (with six members present), it was recorded that "Dr. Storer presented, in the name of Rev. Zadock Thompson, of Burlington, Vt., descriptions and drawings of a new species . . . for which he proposes to establish a new genus under the name of *Salmoperca,* unless it should be found to come under Prof. Agassiz's genus *Percopsis.*"[8]

Next, in the *Proceedings of the Boston Society of Natural History,* William Orville Ayres and Charles Frederic Girard, both accomplished ichthyologists, provided statements of support for Agassiz's genus *Percopsis.* Ayres stated that "on examination of the drawing of the fish called *Salmoperca,* that it evidently belonged to the genus *Percopsis.*"[9] But immediately following the opinion of Ayres appeared a published description of a new species, *Percopsis pellucida,* on behalf of Zadock Thompson.[10]

Although Agassiz' genus *Percopsis* was presented to the Boston Society of Natural History in 1848,[11] his species description was not published until 1850 as *Percopsis gutattus.*[12] *Percopsis* means "perchlike," and *gutattus* means "spotted" or "speckled." Agassiz was particularly fascinated with this fish because it did not fit in his classification system, a system based on scale types that he had published in a multivolume set on fossil fishes (published from 1833 to 1845).[13] At the time, Agassiz thought that fishes with cycloid scales should be classified differently than those with ctenoid scales. Agassiz's ctenoid fish group included a subgroup that he called the percoids, which included perchlike fishes, such as the genus *Perca.*[14] His cycloid fish group included the fish family Salmonidae, which included trout and salmon. Agassiz noted that his new genus *Percopsis* had both ctenoid and cycloid scales, so he assumed that it must be an intermediate between percoids and salmonids. Agassiz wrote:

> The most striking features of the fishes of the tertiary period and those of our time consist in their belonging to two groups of the class only; one, the Ctenoids, with rough, combed scales, in which the respective representatives have also prominent serratures on prominent spines upon the head, in the operculum in particular, and in the fins; the other, the Cycloids, smooth, with simple scales with an entire margin, in which some few types however have also spinous fins.
>
> Now my new genus, *Percopsis,* is just intermediate between Ctenoids and Cycloids; it is, what an ichthyologist, at present, would scarcely think possible, a true intermediate type between Percoids and Salmonidae.
>
> The general form of this genus reminds us of the common perches, but it is easily distinguished from them, by the fact that its

head and the opercular apparatus are smooth and unprovided with denticulations, as also by the presence of a small adipose fin, as in the salmons. The anterior dorsal is also a small fin, composed of soft branched articulated rays, as in the salmons.[15]

It is reasonable to wonder whether Thompson or Agassiz should get credit for the species description. Usually, based on publication date, the person who described the species first is credited with the description. Interestingly, though, Thompson's description in the *Proceedings of the Boston Society of Natural History* did not include his original genus name, *Salmoperca,* but instead used Agassiz's genus name, *Percopsis.* As Poly points out, "Thompson probably submitted his description and drawing to David Humphreys Storer with the name *Salmoperca,* but the published combination is what must be considered."[16] Kendall, assuming that Agassiz published the species name prior to Thompson, lamented that "it is perhaps unfortunate that the nomenclatural rules of priority of publication are so hard and fast that Thompson's name for this fish could not have been retained, in justice to Mr. Thompson."[17] Poly, after further investigation of publication dates, stated that "Thompson's description of *Percopsis pellucida* predates Agassiz's name by one to two months."[18]

This fish tale of two taxonomists has another twist. Neither Thompson nor Agassiz was the first to describe this species, as it was previously described in 1784, when Thomas Pennant, a Welsh naturalist, published the introduction of his book on arctic zoology. Under a section on Hudson Bay fishes, Pennant described a new species, giving it the binomial name *Omisco Maycus.*[19] This name is likely derived from an Algonquian Native American name referencing trout.[20] Based on the description, Pennant's new fish is clearly the same as that described by Thompson and Agassiz.

But the story does not end there. In 1792, the German physician and naturalist Johann Julius Walbaum published a description of the species, citing the 1784 work of Pennant and giving the fish the name *Salmo Omisco Maycus.*[21] This taxonomic tale has a peculiar ending, as the species is currently recognized as the Trout-perch, *Percopsis omiscomaycus* (Walbaum, 1792). Thus, recognition has been given to Thompson's common name "Trout-perch," Agassiz's genus *Percopsis,* Pennant's binomial combined as the specific epithet *omiscomaycus,* and Walbaum as the species authority. Note that Walbaum's name as the species authority is printed in parentheses because the species was originally described as a different genus than *Percopsis.* Based on the code of the International Commission on Zoological Nomenclature (established in 1895), if the genus name has not changed since the original species description, then the authority for the species is not printed in parentheses.

Although *Percopsis omiscomaycus* is neither a trout nor a perch, its common name is apt, as this species shares characters of both trout and perch. Taxonomically, the Trout-perch is placed in the order Percopsiformes and the fish family Percopsidae, and differs from species of trout (order Salmoniformes, family Salmonidae) and species of perch (order Perciformes, family Percidae). There are many ways to look at similarities and differences between species of fish. Without getting into internal anatomy, bone structure, and genetics, a simple approach is to look at the external body characteristics. As previously mentioned, the Trout-perch has an adipose fin, a similar characteristic to species of trout. Relative to body size, trout scales are of extremely small size, whereas scales of perches and the Trout-perch are larger. For example, trout often have more than 110 scales along their lateral line, whereas the number is smaller for Yellow Perch (range 55–64) and Trout-perch (range 43–60). Species of perch have two dorsal fins (a spiny dorsal fin in front of a soft-rayed dorsal fin), whereas species of trout have a single soft-rayed dorsal fin. Trout-perch differ from Yellow Perch and species of trout by having a single dorsal fin with one to two anterior spines followed by soft rays. The head shapes of Trout-perch and Yellow Perch have some similarities, but the margin of the opercle is soft in Trout-perch and sharp-edged in Yellow Perch. Another visual difference is in the position of the pelvic fin. In species of trout, the pelvic fins are generally located abdominally about midway between the pectoral fins and the anal fin. The pelvic fins have a thoracic location (under the pectoral fins) in Yellow Perch, and a subthoracic position (under and slightly behind the pectoral fins) in Trout-perch. Additionally, as Agassiz noted, the Trout-perch has both cycloid and ctenoid scales, whereas trout have cycloid scales and perches have ctenoid scales. An inquisitive naturalist could dissect some specimens and likely find more similarities and differences among these fishes.

Occasionally, someone poses the question, Given the possibility to talk with someone from the past, who would it be? Among the naturalists mentioned in this chapter, would it be Linnaeus, Artedi, Agassiz, or Thompson? I think I would have a sit with Ole Worm, in part to satisfy my curiosity about all the unique pieces and specimens in his cabinet. I would also like to know if he would consider including in his cabinet a specimen of the Trout-perch, a truly unique and interesting small fish of the central Appalachians.

Fragmented Thoughts

I was in a meandering stream of consciousness while driving south on the curvy roads of Route 219 toward Minnehaha Springs, West Virginia. Owing to the central Appalachian geology, driving can be an up- and downhill scenic experience, particularly in the fall season with its flashes of red-and-yellow foliage. Additional complexities of colors are arranged as artwork on barns, often in the form of paintings of large patchwork quilts, providing a reasonable replacement for the yellow-painted words of the previously common Mail Pouch Tobacco signs. There are a lot of interesting things to see while driving along country roads. The complex quilt of the central Appalachian landscape was well characterized by the words of Nathaniel Southgate Shaler (1897), a naturalist and forefather of American geology: "It is true that every hill and valley, every outline of surface, however trifling, indicates some part of the history of that land; but the mass of fact is so great that it is beyond the compass of the mind to grasp it firmly."[1]

Admittedly I am often a distracted driver, not because of a smartphone, but owing to the sights of nature along the roadway. Instead of giving full attention to the road, I am often looking at the landscape, considering the balance of boulders on a hillside slope, watching the trees and wildlife, admiring the beauty of leaves blowing in the wind, and craning my neck to see a creek over a bridge railing. On this drive, I was enjoying my time out of the office, particularly looking forward to wading in a stream, as field work is a naturalist's necessity. My thoughts turned to an oft-cited aphorism of Louis Agassiz, the celebrated Swiss American naturalist, zoologist, and geologist: "The pupil studies nature in the classroom, and when he goes out of doors he cannot find her."[2] On this day, I was outside marveling at nature in all of her glory.

During the drive, my thoughts were free flowing on ideas for a chapter of this book, when my mind wandered to Henry Wadsworth Longfellow, the well-known American poet. In 1855, Longfellow published *The Song of Hiawatha*, which included the fictional character Minnehaha, a Native American princess,[3] whose name happened to be the inspiration of my destination. My plan was to drive to Minnehaha Springs, park the old suburban in an "ichthyological pulloff" along Knapp Creek, and seine fishes from the stream. An ichthyological pulloff is a wide spot sandwiched between road and stream, providing a good place to park. Often a pulloff is next to a bridge, which leads to the absurd claim by some that ichthyology is "the study of fishes that live under bridges."

After pulling over, with seine and minnow bucket in hand I climbed down the bank under the bridge support. My thoughts were on the diversity of fishes in Knapp Creek, and particularly on which species I might capture with the seine. Two species of chubs known to occur in this creek (Bigmouth Chub and Bluehead Chub) took my thoughts back to Longfellow, because their genus, *Nocomis,* was named after Hiawatha's grandmother, Nokomis. My purpose for sampling Knapp Creek was to collect and photograph two species of sculpin (Mottled Sculpin and Kanawha Sculpin) for a book on West Virginia fishes. Sculpins are small fishes, usually less than three inches (7.6 cm) in length, with wide, somewhat flattened heads and large pectoral fins. Some people commonly call sculpins "Miller's Thumb," whereas the fly angler's mimic is called the "Muddler Minnow."

I left Knapp Creek (with some decent photographs) on a different route, southeast on Route 39. With my meandering mind, I remembered that Longfellow also wrote a poem about a well-known scientist, one whose research focused on many areas, including the study of glaciers and the study of fishes. One stanza from this poem is as follows:

> Though at times he hears in his dreams
> The Ranz des Vaches of old,
> And the rush of mountain streams
> From glaciers clear and cold;[4]

A ranz des vaches is a traditional Swiss melody played with a horn by an Alpine herdsman driving his cattle. The poem is about Louis Agassiz, who was born in Switzerland in 1807 and died in Massachusetts in 1873. Agassiz was a professor at Harvard University and was one of the first scientists to propose that earth had experienced previous ice ages. He was a student of glaciers, but also had a fondness for fishes, including sculpins.

Like my thoughts, nature is often quite fragmented, continuity and connectedness are not requirements, and unforeseen relationships are often found fortuitously. As I drove, my connected stream of fragmented thoughts on Shaler,

FIGURE 14.1. A Checkered Sculpin with a length of 2.7 inches (6.9 cm).

Agassiz, Longfellow, and sculpins ultimately led me to consider a chapter on the Checkered Sculpin. This fish was named for its appearance—individuals characteristically have a checkered series of pigment blotches on their lower sides.[5] Although this fish has a common name, the Checkered Sculpin is still undescribed, meaning that it does not have a scientific name (figure 14.1). Previously, the Checkered Sculpin was considered a Slimy Sculpin (*Cottus cognatus*), in part because of the shared character of having three fin rays in each pelvic fin. All other described species of *Cottus* in the central Appalachians have four rays per pelvic fin: Mottled Sculpin, Blue Ridge Sculpin, Potomac Sculpin, and Kanawha Sculpin. The current name representing the Checkered Sculpin's undescribed status is *Cottus* sp. cf. *cognatus,* meaning a species (sp.) of the genus *Cottus* that "confers" or "compares to" the Slimy Sculpin (cf. is an abbreviation for the Latin word *conferre*).

The Checkered Sculpin lives in the upper Potomac River drainage within the Ridge and Valley province of Maryland, Pennsylvania, Virginia, and West Virginia (figure 14.2). Within the Ridge and Valley province, many of the populations are within the Shenandoah and Cumberland Valleys of the Great Appalachian Valley. Interestingly, populations of the Checkered Sculpin are almost always associated with high-volume cold-water springs, and are often isolated from each other by large geographic distances. Populations of Checkered Sculpin are known from the Antietam Creek catchment area of Maryland and Pennsylvania, including Little Antietam, Beaver, and Marsh Creeks. The Conococheague Creek catchment of Maryland and Pennsylvania supports populations in Muddy Run and Falling Spring Branch. This fish also occurs in tributaries of the Shenandoah River watershed of West Virginia (Bullskin and Long Marsh Runs) and Virginia (Pass Run, Middle River, and Mossy and Whisky Creeks). Populations are found at many spring habitats in the Opequon Creek watershed of West Virginia and Virginia, including Tuscarora and Mill Creeks and Evans, Hopewell, and Turkey Runs. Additionally, West Virginia populations occur in Harlan and Rocky Marsh Runs. To the west of the Great Valley, populations are known from the Cacapon

River drainage (Edwards and Trout Runs) and the headwaters of the South Branch of the Potomac River (Hammer Run, Thorn Creek, and Strait Creek). To the east of the Great Valley is the carbonate karst region of the Frederick Valley,[6] which supports at least one Checkered Sculpin population in Carroll Creek near Frederick, Maryland. An adventurous person could easily find additional populations of the Checkered Sculpin in the karst areas of the Ridge and Valley section of the upper Potomac River drainage, as one only needs to locate the cold-water high-volume springs. As with all species of nature, there are a lot of intriguing questions to ask, but one particularly stands out: why are the locations of Checkered Sculpin populations so predictably positioned at or near high-volume cold-water springs?

The Checkered Sculpin was the subject of my master's thesis,[7] so I spent a few years thinking a lot about this fish. The thought of focusing solely on a single fish for such a long time reminds me of stories by Louis Agassiz's students. One, Nathaniel Southgate Shaler, was sitting at a small pine table, where Agassiz placed a small fish in a rusty tin pan. According to Shaler, Agassiz required that "I should study it, but should on no account talk to any one concerning it, nor read anything relating to fishes, until I had his permission so to do. To my inquiry 'What shall I do?' he said in effect: 'Find out what you can without damaging the specimen; when I think that you have done the work I will question you.'" After a week of studying the specimen, Shaler was given the opportunity to report his findings. Agassiz responded with "That is not right." Shaler continued:

> I went at the task anew, discarded my first notes, and in another week of ten hours a day labor I had results which astonished myself and satisfied him. Still there was no trace of praise in words or manner. He signified that it would do by placing before me about a half a peck of bones, telling me to see what I could make of them, with no further directions to guide me. . . . Two months or more went to this task with no other help than an occasional looking over my grouping with the stereotyped remark: "That is not right." Finally, the task was done and I was again set upon alcoholic specimens,—this time a remarkable lot of specimens representing, perhaps, twenty species.[8]

Shaler then used Agassiz's system of classification, which was based on fish scale types. Interestingly, he found one specimen that had one scale type on one side and a different scale type on the other side. This finding surprised Shaler:

> This not only shocked my sense of the value of classification in a way that permitted of no full recovery of my original respect for the

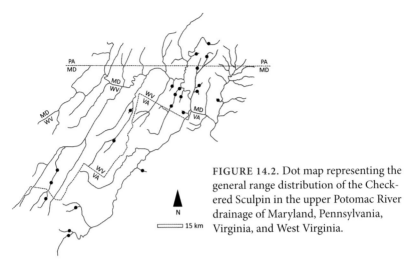

FIGURE 14.2. Dot map representing the general range distribution of the Checkered Sculpin in the upper Potomac River drainage of Maryland, Pennsylvania, Virginia, and West Virginia.

N

15 km

process, but for a time shook my confidence in my master's knowledge. At the same time I had a malicious pleasure in exhibiting my find to him, expecting to repay in part the humiliation which he had evidently tried to inflict on my conceit. To my question as to how the nondescript should be classified he said: "My boy, there are now two of us who know that."[9]

In some ways I shared an experience similar to that of Shaler, as I spent hundreds of hours in my master's research in microscopic examination of over 500 Checkered Sculpins from populations across its range.[10] My professor, Richard L. Raesly, had previously conducted an extensive study on the Checkered Sculpin and thus had a general expectation for my final research findings. As a student, my role was to develop skills in scientific inquiry, although in doing so I learned a great deal about the morphology of this fish. In the end, there were two of us who knew a lot about the Checkered Sculpin. In the words of Shaler: "I had learned the art of comparing objects, which is the basis of the naturalist's work."[11]

Continuing my drive, I proceeded southeast from Minnehaha Springs. I passed through Warm Springs, Virginia, headed southward to the town of Hot Springs for a quick lunch, then reversed direction, weaving my way to Millboro Springs and Augusta Springs. Ultimately, I reached Route 340 for a northward drive down the Shenandoah Valley, a stronghold for both high-volume spring-fed streams and populations of Checkered Sculpins. The length of this magnificent valley is about 140 miles (225.3 km), bordered by the Blue Ridge Mountains (southeast) and the North and Shenandoah Mountains (northwest).[12] From driving this route many times before, I enjoy the familiar landmarks and scenic views, but the sinkholes on the landscape always fascinate me anew. Sinkholes, often as bowl-shaped depressions on the earth's surface, are present because of the karst landscape, a topography underlain by limestone. The sinkholes form by acidic dissolution of the limestone bedrock, often involving the collapse of underground limestone caverns.

Underground drainage systems are common within fractures of carbonate (limestone) rocks of karst landscapes, then often emerge on earth's surface as springs. These springs provide unique habitats for aquatic life, where the temperature of the spring water is usually stable across seasons. Warm springs have relatively high water temperatures, such as the 96°F (35.5°C) spring pools at Warm Springs, Virginia. Other springs have higher water temperatures, such as the 104°F (40°C) Boiler Spring at Hot Springs, Virginia.[13] Warm springs have water temperatures of at least 9°F (5°C) higher than mean annual air temperatures, whereas hot springs have water temperatures that exceed 100°F (37.8°C).[14] The Ridge and Valley province of the central Appalachians has the highest concentration of thermal springs in the eastern United States.[15]

High-volume cold-water springs within the Potomac River drainage of the Ridge and Valley province are often located on or near faults.[16] The faults provide large conduits for vertical water flow and ultimately high-volume spring outflow. Within this region, streams with high-volume cold-water springs are often the locations with stable populations of the Checkered Sculpin. For example, Priest Spring of the Opequon Creek drainage supports a localized Checkered Sculpin population and has water temperatures at about 53.0°F (11.7°C) and a discharge of 6,421 gallons per minute (gpm), or about 35,000 cubic meters per day (m³/d).[17] Tuscarora Creek, a tributary of Opequon Creek, has a least 13 contributing springs, including Kilmer Spring, with water temperatures of 54.1°F–55.6°F (12.3°C–13.1°C) and a range of discharge rates from 2,568–2,899 gpm or 14,000–15,800 m³/d.[18] Many springs occur within the Antietam Creek basin, including Powell Spring on Beaver Creek (north of Boonsboro, Maryland), with water temperatures of 53.6°F–55.9°F (11.9°C–13.3°C)[19] and a discharge of at least 3,000 gpm or 16,353 m³/d.[20] Falling Spring of the Conococheague Creek catchment, about three miles (4.8 km) east of Chambersburg, Pennsylvania, has a discharge of about 1,880 gpm or 10,248 m³/d.[21] Harlan Run, a direct tributary of the Potomac River, is influenced by at least seven springs,[22] with Harlan Spring (52.2°F, 11.2°C) discharging about 2,165 gpm or 11,800 m³/d.[23]

During the Pleistocene and Holocene Eras, the climate of North America has undergone dramatic swings, where temperatures have cycled between colder and warmer periods (see chapter 12, "Trees and Trout"). Within the central Appalachians, previous periods of colder climates are generally associated with periods of glacial advance. During previous cold periods, cold-water fishes likely had wider distributions within the central Appalachians.

Examples of cold-water fishes, that is, species that are intolerant to warm water, include Brook Trout, Allegheny Pearl Dace, Redside Dace, and Longnose Sucker. These species have experienced habitat fragmentation, particularly on the southern extents of their ranges, where contiguous cold-water habitats have been fragmented. Habitat fragmentation results in range fragmentation, where contiguous ranges are reduced to isolated, localized populations.

The Checkered Sculpin was likely distributed widely within the upper Potomac River drainage during periods of colder climate. With a warming climate and increasing stream-water temperatures, the distribution range of the Checkered Sculpin contracted, resulting in a fragmented range with populations associated with cold-water spring habitats. During initial stages of climate warming, the Checkered Sculpin was likely present at low-volume cold-water springs, but these habitats are more vulnerable to environmental variation and droughts than those of high-volume springs. From a conservation perspective, the extirpation of each population isolate takes the species one step closer to extinction, a process called the "extinction ratchet."[24] The contraction and expansion of populations likely repeated multiple times. Currently, the summer water temperatures of most streams within the upper Potomac River drainage are too warm for Checkered Sculpins, thus the range is highly fragmented with geographically separate populations found at or near high-volume cold-water springs.

As I close this chapter, I'll offer my own poem on the Checkered Sculpin, written in Longfellow's trochaic tetrameter (rhythm of each line in four feet, Dum da, Dum da, Dum da, Dum da). Each metrical foot is referred to as a trochee, where a stressed syllable is followed by an unstressed one. An example line of four feet, with the accented syllable of each trochee in bold, would be **Pop**u**la**tion **frag**men**ta**tion, which can also serve as the title to my poem:

> With the current cycle warming,
> Species ranges often changing;
> Sculpin ranges not expanding,
> Checkered Sculpin range reducing;
> Range reduction, fragmentation
> Cold springs linked to distribution.

With the contributions of Agassiz, Shaler, and Longfellow, it is clear that geology and poetry influence ichthyology, but I am not as sure that verse from an ichthyologist (at least from yours truly) will shape the fields of geology and poetry. Given that you've read my attempt at poetic prose, there are now two of us who know that.

Satellite Sunfish

Each year, during late spring to early summer, I find myself standing on a lakeshore, staring intently into the water. The lake bottom, which looks a lot like a lunar landscape, is covered with clusters of small craters created by colony-nesting sunfishes. One of my favorite sunfishes is the Bluegill (*Lepomis macrochirus*), which occurs throughout the central Appalachians, frequently found in rivers, reservoirs, and ponds. The name *macrochirus* means "large hand," probably in reference to the long pectoral fins. It takes a large hand to hold a large Bluegill, as this sunfish has a deep body relatively to its length, a body shape that supports the generic sunfish name of "panfish." The Bluegill is a beautiful fish (figure 15.1). Brilliantly colored large breeding males, often reaching nine to ten inches (22.9–25.4 cm) in length, have a bright orange breast, blue-green sides, and bright blue stripes on the head. These large males reach sexual maturity at about age seven or eight. Females, which often mature at about age four, and most smaller males are not as colorful.

I have an unusual fascination with fish reproductive behavior. My therapy is to talk about it on a regular basis, but then I often get asked the question, how can spawning fish really be all that fascinating? An example may be the best answer, so here's an explanative essay on the reproductive ecology of the Bluegill. The complexity of colony-nesting Bluegills is a marvel, one that will boggle your naturalist mind.

Bluegill breeding behavior is well documented. After establishing a territory, a large male creates a circular nest. He fans or sweeps the bottom with his caudal fin, creating a craterlike or bowl-shaped depression. Nest construction often takes a few days to complete. The nest-constructing male, also known as a parental male, gets vertical, perpendicular to the bottom,

FIGURE 15.1. A Bluegill Sunfish approximately nine inches (22.9 cm) in length from a Central Appalachian reservoir.

during tail-sweeping. Other large parental males repeat this process nearby, collectively creating a cluster of nests. The number of nesting males in a colony varies widely, but can exceed over 70. Nests in the colony are sometimes less than four inches (10.2 cm) apart. Spatial position of nests can also be influenced by weedy areas on the spawning grounds. Parental males guard and protect the nest and proximate surroundings, a territorial behavior limited in range by nearby nesting males.

While waiting for a female's approach, a nesting male may tail-sweep the bottom of his nest, or swim around the perimeter, a behavior known as rim circling. The male encircles an approaching female, leading her to the nest. Occasionally, a parental male spawns simultaneously with two females. Vernon L. Avila described the spawning sequence:

> Once the female is settled in the nest the male swims close beside her. They face in the same direction and circle together in the nest, with the male on the outside and the female inside. During the spawning act the male's anterior ventral surface darkens. As they circle in the nest the male remains upright, but the female tilts her dorsum away

from the male and, leaning to one side, rubs the ventral surface near her genital pore in a quick "vibrating" movement against the side of the male near his anal fin. As they circle together, the female tilts into this spawning posture every 60–100 sec. As contact is made, eggs and presumably milt are deposited (Miller, 1963). Spawning sessions varied in length from 15–90 min.[1]

The "female tilt" described by Avila is also known as the "female dip."[2] The spawning male must be able to recognize this important female behavior, because when the female dips to deposit eggs, the male must be closely paired with the female to fertilize the eggs.

Often in nature parental care is provided by the mother, but this is not the case in sunfishes. Female Bluegills are promiscuous, often spawning in the nest of more than one parental male. Also, a male courts and spawns with multiple females, accumulating as many as 30,000 eggs in his nest.[3] Following spawning, the parental male remains with the nest for about ten days, depending on water temperature, initially protecting the eggs and then guarding the young. The eggs are fanned by the male for two to three days or until hatching. Young are protected until their departure in about four to seven days. Within a breeding season, colony nesting occurs across several time periods.[4] The largest parental males often form colonies and spawn earlier than smaller parental males, and a male may spawn at two or more time periods within the breeding season.[5]

Colony nesting is not unusual in nature, but is not a typical strategy for most fishes of the central Appalachians. Within a colony, it would be reasonable to expect extreme territorial aggression among closely positioned nesting Bluegills, an interaction known as the "nasty neighbor effect."[6] Instead, observations on colony-nesting Bluegills have found that, once nesting territories are established, nearest neighbors become increasingly cordial.[7] A territorial truce is often made between "familiar" nesting neighbors, but with continued concern and suspicion. Nesting males continue to be aggressive toward intruding strangers. This behavior, now known as the "dear enemy effect,"[8] was initially proposed in 1954 by the British naturalist and ornithologist James Maxwell McConnell Fisher: "The effect is to create 'neighbourhoods' of individuals which are masters of their own definite and limited property, but which are bound firmly, and socially, to their next door neighbours by what in human terms would be described as a dear enemy or rival friend situation."[9]

Nest building and colony nesting are interesting parts of sunfish reproduction, providing the background for an even more interesting story about individual reproductive tactics. The tale of two spawning sunfish has a twist: sometimes a threesome occurs in the nest, where a third individual joins in with the sexual act of the parental male and female. Early research on sunfish spawning reported that a second female often joined a breeding pair in the nest.[10] More recently, from a field study of Bluegill nesting behavior, Avila noted: "In three instances there were more than one male and female spawning in a nest. While a pair of bluegills were spawning, another female would slip in and align herself with the male or in some instances align herself with the other female. The intruding female would tilt herself into the spawning posture with either the male or female."[11]

The plot thickens, however, as more recent research has realized that the third sexual partner is often not a female after all, but a female mimic, a cross-dressing male that takes on the appearance and behavior of a female.[12] The female mimic is also called a satellite male. In other cases, the third partner is not a female mimic, but a smaller male that "sneaks" in on the spawning parental male and female.[13] The female mimics and sneaker males do not construct nests and do not partake in parental care of eggs or young. Thus, the parental male, who provides sole parental care, is not the sole sire of eggs in his nest. Instead, the large male is cuckolded by these smaller males, because he is tasked with parental care of their eggs and young. Most sneaker males spawn at age two, but can range from ages one to five. Sneaker males convert to female mimics, often at age four, but female mimics can range from ages three to six. Cuckolder males, which never become parental males, die at about age six.[14] Parental males wait until about age seven or eight to mature and spawn, and generally do not exceed age 11.[15] The parental male and the cuckolder male are alternative life history tactics, where both are successful at fertilizing eggs. But which tactic is more successful, the parental male or the cuckolder male?

It would be interesting to know the proportion of cuckolder males to parental males in a population, a ratio that would presumably also be represented in each spawning colony. One study found that the proportion of sexually mature males within a population is about six cuckolder males to one parental male.[16] It would be reasonable to assume that a higher density of cuckolder males in the population should result in higher rates of cuckolder spawning success. This assumption, however, does not account for the aggressive nature of parental males toward preventing the spawning success of cuckolder males, as outlined by Mart R. Gross:

In a study of pairing success at seven colonies (Gross 1982), cuckolders attempted to intrude into the nests during nearly 60% of the female dips. Most of these attempts were blocked by parental males, and only 14% of all female dips were successfully paired by cuckolder males. Females did not avoid cuckolder males, but rather spawned readily with them even while the parental male was chasing egg predators or other cuckolders. Thus the primary determinant of cuckolder pairing success was avoiding the parental male. This ability varied among colonies, and was affected by ecological factors as well as interactions among cuckolders themselves. As a consequence, the pairing success at the seven colonies ranged from about 3% to 34% of female dips.[17]

An aggressive parental male actively chases cuckolders from his nest, but this only represents part of the story on the cuckolder's constraints. Gross also mentions the importance of ecological factors and interactions among cuckolders.[18] An important ecological factor is the presence of aquatic vegetation (weeds) on the colony grounds. Weeds provide cover for sneaky cuckolder males, affording protection from predators and parental male aggression, often resulting in a higher density, as number of individuals per area, of cuckolder males.

Interactions among cuckolders differ depending on their density, and can result in both increased and decreased spawning success. In describing the relationship between cuckolder interactions, and pairing success at two colonies, Gross wrote:

> Cuckolder pairing success was density dependent within each colony. . . . However, the density providing peak average success per cuckolder was negatively correlated with the amount of cover at the colony. . . . Thus in colony A, with a high of 88% cover, average pairing success per cuckolder peaked at 17% of female dips when only a single cuckolder was present at the nest site and declined to 4% of dips with 11 cuckolders present. By contrast in colony B, with 57% cover, there was an initial increase in average pairing success with density. Here, a single cuckolder achieved 18% of the female's dips, but two cuckolders at the nest each achieved 27%. Success then declined to near 7% with 10 cuckolders present.[19]

A single cuckolder at a nest is often unsuccessful at pairing with the female because he is continuously chased away by the parental male. When two to four cuckolders are present at a nest, one of the cuckolders may enjoy spawning success because the parental male chases the others away from the nest. When as many as eight or more cuckolders are present, then spawning success is reduced because cuckolders are competing with themselves and the parental male. Often, in the presence of a high number of cuckolders, the parental male's aggression is increased to the point where the female leaves the nest.

Parental males may further reduce spawning success of cuckolder males by selectively cannibalizing offspring.[20] Studies have demonstrated that a parental male uses olfactory cues (smell) to determine which young offspring are his or from a cuckolder.[21]

The alternative reproductive tactics (parental vs. cuckolder) of the male Bluegill have been considered as a conditional strategy.[22] An individual likely makes a choice, based on its body-condition state, as to whether it will pursue cuckoldry at a young age or wait until later in life to become a parental male. Alternative-state conditions are likely to be small versus large body size at a young age, where the larger-bodied individuals will likely choose the parental tactic. Conditional states, however, do have a component of genetic inheritance, as individuals that attain a large size at a young age may also have offspring with similar growth patterns.

Parental and cuckolder tactics do not often have equal fitnesses, where fitness is defined as the "degree of demographic difference" between the two tactics.[23] As previously mentioned, the number of cuckolder males in the population exceeds that of parental males. This fitness tradeoff was explained by Gross and Joe Repka:

> The fitness of the strategy is at the population level while the fitness obtained by each individual is through the expression of its tactics. The fitness of the strategy is not determined by either tactic alone but by how the strategy allocates individuals in the population to the two tactics. Each tactic contributes to fitness but their average fitnesses are unlikely to be equal. The overall fitness of the strategy is maximized through the appropriate allocation of tactics such that each individual does the best that it can given its state. This of course also benefits the individual.[24]

Depending on the situation, both reproductive tactics can be successful. It is important to consider, however, that the cuckolder tactic is only possible because of the parental male's care of the offspring. Without parental care, the eggs and fry would not likely survive nest predation.

Now, given your new knowledge about Bluegills, your inner naturalist is fully aware of this fascinating and reproductively complex central Appalachian fish. After reading this chapter, you may find that people are starting to talk—particularly talking about your seemingly deranged behavior of standing on a lakeshore, staring intently into the water. Not to worry, however, as talking about it with friends and colleagues is your best therapy.

Gamest Fish That Swims

I have sincere admiration for the Smallmouth Bass, a species of fish that lives in many of central Appalachia's creeks, rivers, and reservoirs. Hereafter referred to as Smallmouth, it is native to most of the Ohio River catchment area of the central Appalachians. It is introduced to all Atlantic slope drainages of the region. An introduced fish population, also known as a nonnative population, is one that has become established in a watershed outside of its native range. Most nonnative populations of Smallmouth result from people moving this species across drainage divides. In the central Appalachians, many of the introduced Smallmouth populations date back to the mid to late 1800s. During this time, fish populations were dwindling from habitat loss and overharvest, causing agencies or private citizens to turn to stocking fishes as an answer for improving opportunities for recreational and subsistence fishing.

At a young age, my introduction to the Smallmouth was through angling, a lifelong passion that initially nurtured my interest in nature. People experience nature in many ways, but angling continues to be one of my main conduits to it. I continue to be a strong advocate for Smallmouth fishing. The naturalist and avid angler James Alexander Henshall was one of the first to promote this species as a sport fish. In his *Book of the Black Bass,* published in 1881, Henshall stated that the Smallmouth is "*inch for inch* and *pound for pound,* the gamest fish that swims."[1] A game fish or sport fish is a fish commonly targeted by anglers. In Henshall's time, Black Bass was a name used for two species of fish in the central Appalachians: Smallmouth and Largemouth Bass (figure 16.1). Currently, the term applies to three species in the central Appalachians: Smallmouth, Largemouth, and Spotted Bass.

FIGURE 16.1. Photographs of Smallmouth Bass (*top*), Largemouth Bass (*center*), and Spotted Bass (*bottom*).

In 1802, Smallmouth and Largemouth Basses were described as two species new to science by the French naturalist Bernard Germain de Lacépède.[2] The scientific names he gave them were *Micropterus dolomieu* and *Labrus salmoides*. Subsequently, numerous new names for these two species were suggested, many by well-known naturalists, including Constantine Samuel Rafinesque, Charles Alexandre LeSueur, Georges Cuvier, and James Ellsworth DeKay. To add to the confusion, the same scientific name was sometimes used for both species. Even more disconcerting, the names of these two species were commonly switched in the published literature, where the name *salmoides* was often associated with the Smallmouth. It seems odd that so much taxonomic uncertainty was attached to these two species. Henshall, summarizing the dissatisfactory taxonomic history of this fish, lamented:

> Lacépède had another specimen of the Black Bass, without label, and from an unknown locality. This one had the last rays of the dorsal broken and torn loose from the rest, and was otherwise in a forlorn condition. This specimen he considered as a genus distinct from the other, and he gave it the name of *Micropterus dolomieu*—"Dolomieu's small-fin." Dolomieu was a friend of Lacépède, who had had about as much to do with the fish as George Washington or Victor Hugo. No one could tell, either from figure or description, what this *Micropterus dolomieu* was; but Cuvier, thirty years later, found the original type and pronounced it a Black Bass, in poor condition, and declared that "the genus and species of *Micropterus* ought to disappear from the catalogue of fishes." Then the versatile and eccentric Professor Rafinesque appeared upon the scene, and in rapid succession gave the small-mouthed Black Bass names enough for a whole family. First he called it *Bodianus achigan*, being told that the Canadian voyageurs knew the fish as *L' achigan*. Then afterward specimens of different sizes appeared as *Calliurus punctulatus, Lepomis trifasiata, Lepomis flexuolaris, Lepomis salmonea, Lepomis notata,* and *Etheostoma calliura*. Soon after Le Sueur, with a lofty scorn for Rafinesque and his doings, named specimens of different sizes, *Cichla fasciata, Cichla ohiensis,* and *Cichla minima*. Lastly, DeKay, in 1842, called it *Centrarchus obscurus,* and we hope this may be the last.[3]

The scientific names for the two species were stabilized as *Micropterus dolomieu* and *Micropterus salmoides* by the prominent ichthyologists David Starr Jordan and Charles Henry Gilbert in their 1882 publication *Synopsis of Fishes of North America*.[4] This decision was made based on recommendation by Henshall. As a naturalist, Henshall was a proponent of the original scientific names, but emphasized that both species should be considered as members of the genus *Micropterus*. In lengthy prose, Henshall wrote:

> I am aware that the change made in the scientific names of the Black Bass species, will be looked upon by many, at first, as unwise and injudicious; but, under the circumstances, I could not have done otherwise than to restore the names rightfully belonging to them, inasmuch as by so doing it clears up the former confusion attending the

nomenclature of the species, and renders plain the causes of the same. Moreover, as the names *Micropterus* for the genus, and *dolomieu* and *salmoides* for the species, are the first ever bestowed, in each instance. their adoption will be not only an act of justice, but of expediency, for as we cannot go back of them, it sets the matter at rest, forever.[5]

I find myself agreeing with Henshall, at least on three commonalities: a need for nature, a fever for fishing, and clarity of taxonomic nomenclature. As an angler and author, Henshall was a strong advocate for Smallmouth fishing. Henshall's bass passion was accurately reported by Robert E. Jenkins and Noel M. Burkhead,[6] who noted that "the advocacy of the Smallmouth by J. A. Henshall truly endeared it to the American angler." Henshall wrote: "The Black Bass is eminently an American fish, and has been said to be representative in his characteristics. He has the faculty of asserting himself and making himself completely at home wherever placed. He is plucky, game, brave and unyielding to the last when hooked. He has the arrowy rush and vigor of the Trout, the untiring strength and bold leap of the Salmon, while he has a system of fighting tactics peculiarly his own."[7]

To use Henshall's words, the Smallmouth has certainly "made himself at home" as a nonnative species in the central Appalachian drainages of the Susquehanna, Delaware, Potomac, James, and New Rivers. Why was the Smallmouth so successful at populating these river systems outside of its native range?

Dr. John Eoff of Wheeling, West Virginia, documented perhaps the first introduction event of Smallmouth in the Potomac River drainage, where fish were transferred from Wheeling Creek, West Virginia, in 1854:

> Mr. William Shriver, a gentleman of this place, and son of the late David Shriver, esq., of Cumberland, Maryland, thinking the Potomac river admirably suited to the cultivation of the bass, has commenced the laudable undertaking of stocking that river with them; he has already taken, this last season, some twenty or more in a live box, in the water tank on the locomotive, and placed them in the canal basin at Cumberland, where we are in hopes they will expand and do well, and be a nucleus from which the stock will soon spread.[8]

It seems remarkable that 20 Smallmouth in 1854 could survive the 211-mile (340-km) train ride along the Baltimore and Ohio Railway from Wheeling, West Virginia, to Cumberland, Maryland. Within 20 years following the initial introduction event, and likely aided by additional undocumented introduction events, the Smallmouth population rapidly expanded its range within the Potomac River drainage. At this time, people of Maryland were already considering this species an equal to the native fishes. In 1876, the naturalists Phillip Reese Uhler and Otto Lugger wrote:

> Though of recent introduction into our waters, it has become so plentiful, and plays so important a part in the food supply of the interior of the State, that we refer to it among the native fishes. The black bass has not, nor is it likely that it will be so constantly found in our markets, as those fish which find their subsistence, and obtain their growth in the sea, where they find an inexhaustible supply of excellent food. But the bass fills, and well fills a long felt deficiency of fish food in the interior fresh water, occupying the middle ground between the rapid leaping brook, the home of the speckled trout (*Salmo fontinalis*) and that frequented by the white and yellow perch. They furnish wholesome fish diet, and recreation to those who before their introduction had only the Cat, the Sucker, the Fall fish and a few minnows.[9]

Uhler and Lugger noted that the Smallmouth introduced in the Potomac filled a geographical gap between the headwater fishery of speckled trout *Salmo fontinalis* (now known as Brook Trout *Salvelinus fontinalis*) and the near-coastal fishery of White Perch and Yellow Perch.[10] Thus, Henshall was not the only person in the late 1800s with an interest in Smallmouth, although he was perhaps the predominant advocate for the species at the time. Henshall recognized the popularity of trout fishing, but suggested that the Smallmouth would soon be the premier game fish in the region (figure 16.2).[11]

Henshall did not note potential problems for native fishes relative to the introduction of predatory Smallmouth, but he did recognize that nonnative predators could negatively impact Brook Trout populations: "That he will eventually become the leading game fish of America is my oft-expressed opinion and firm belief. This result, I think, is inevitable; if for no other reasons, from a force of circumstances occasioned by climatic conditions and the operation of immutable natural laws, such as the gradual drying up, and dwindling away of the small Trout streams, and the consequent decrease of Brook Trout, both in quality and quantity; and by the introduction of predatory fish in waters where the Trout still exists."[12]

Today few people know that Smallmouth are not native to the Potomac River drainage, and perhaps fewer know that many other populations of this species in the central Appalachians are also nonnative. The Susquehanna and Delaware Rivers' populations of Smallmouth are nonnative. In

FIGURE 16.2. A 1905 photograph of two anglers and their catch of Smallmouth Bass from the South Branch of the Potomac River at Petersburg Gap, West Virginia. Photo adapted from Clark (1980), courtesy of the West Virginia Division of Natural Resources.

1874, Professor James Wood Milner, an employee of the US Fish Commission, indicated that Pennsylvania citizens introduced Smallmouth into the Susquehanna River at Harrisburg, Pennsylvania, around the year 1869, and further stated that "in 1873 the tributaries of the Susquehanna, the Potomac, and Delaware Rivers were supplied with black bass by the commissioners at thirty-five different points."[13] In 1892, Tarleton Hoffman Bean, an ichthyologist from Pennsylvania, noted that 450 individuals in 1870 were transferred from Harpers Ferry (Potomac River drainage) to the Delaware River below Lehigh Dam at Easton, Pennsylvania.[14] Furthermore, in 1893, William Edward Meehan, associate editor of the *Philadelphia Public Ledger*, noted that shortly after the 1870 introduction a number of "public-spirited citizens" purchased additional fish from Harpers Ferry for transfer to the Susquehanna River basin and the Schuylkill River of the Delaware basin:

> The result surpassed their expectations. The fish took kindly to their new quarters and multiplied in such amazing quantity that in three years they were caught in the Delaware, Susquehanna and Schuylkill rivers in great numbers. Fish four and five pounds in weight were frequently caught in 1873. The voracity and eagerness with which

they took both bait and fly, the stubbornness and vigor with which they fought for freedom and life when hooked, speedily made them a favorite game-fish among anglers, many old brethren of the rod, indeed, declaring that the new fish yielded more sport than the speckled trout.[15]

Meehan gave information on the population status of Smallmouth introductions in Pennsylvania: "So rapidly did the bass multiply in the Delaware river that three years after their the fish commissioners were able to stock other waters therefrom, and in 1873 no less than two thousand and forty-four were captured near Easton and distributed throughout the State. These bass were deposited in the Lehigh river, the North and West branches of the Susquehanna, the Juniata and its branches, the reservoir near Hollidaysburg, Yellow Breeches creek, Pequa creek, Chiquesalonga creek, Octorora creek and Codoras creek."[16]

The James River watershed of Virginia and West Virginia, with upper sections in the central Appalachians, is another Atlantic slope drainage with another nonnative population of Smallmouth. In 1877, a Virginia Fish Commission report indicated that Smallmouth were introduced to the James River drainage in about 1871. Introduction events were at various locations between the Jackson River at Covington, Virginia, and downstream to the mouth of Tye River at Norwood, Virginia. As with other introductions of this fish in the region, the Virginia Fish Commission stated that Smallmouth had "become quite numerous" within about five years following this stocking event.[17]

After considering the introduction events of Smallmouth in sections of the Susquehanna, Delaware, Potomac, and James Rivers' catchment areas of the central Appalachians, it is interesting to note that this species was also introduced into the New River drainage. Although the New River is an upper watershed of the Ohio River drainage, the Smallmouth and many other fishes native to the Ohio River drainage were historically absent from the New River owing to the barrier of Kanawha Falls at Glen Ferris, West Virginia. The New and Gauley Rivers join to form the Kanawha River a short distance upstream of Kanawha Falls. The diversity of the native fish fauna above Kanawha Falls is less than below the falls. Edward Drinker Cope, well respected for his scientific contributions to paleontology, herpetology, and ichthyology, surveyed the fish fauna of the New River catchment in 1867.[18] Cope did not find Smallmouth to be present in the New River catchment in 1867, providing supporting evidence for nonnative status of the current population. Further, the Virginia Fish Commission, in a published statement about game fishes, wrote that the "New river, which rises in North Carolina and traverses the counties of Carroll, Wythe, Pulaski and Giles, has in its main stream little else than the cat fish."[19] Channel Catfish, Flathead Catfish, Green Sunfish, American Eel, and Brook Trout are the only native predatory fishes of moderate to large size in the New River drainage.[20]

Many undocumented introduction events likely occurred during the late 1800s, as Milner wrote in 1874: "There have been very many transfers of these valuable species that have not been recorded, as they are easily kept alive while being moved from one place to another, and propagate surely and rapidly in ponds, lakes, and rivers. These details are given because they show the facility with which comparatively barren waters may be stocked to a considerable extent with good food-fishes, and they exhibit the general interest and attention that have been given to this mode of propagation."[21]

As I've described, the fish faunas of the central Appalachian sections of the Atlantic slope and New River drainages were historically depauperate of mid- to large-sized predatory fishes. Milner clearly recognized this by stating that the success of introduced Smallmouth populations "show the facility with which comparatively barren waters may be stocked to a considerable extent."[22] Milner's reference to "barren waters" emphasized the lack of game fishes (i.e., predatory fishes) in the rivers of the central Appalachians. Without large predators, the introduced Smallmouth had little competition for food, allowing it to exploit the available prey resources. As previously emphasized by Uhler and Lugger, the Virginia Fish Commission, and Meehan, nonnative Smallmouth populations in the central Appalachians rapidly expanded their ranges and became noticeably abundant in a short time after initial introduction events. An abundance of prey species, such as minnows, crayfishes, and aquatic insect larvae, likely supported the fast growth of individual bass and the rapid expansion of these introduced populations.

This chapter may appear to promote a positive spin on the introduction of fishes outside of their native ranges. Not so. My hope is that you will find this story as an interesting history of Smallmouth in the central Appalachians. As a naturalist, I have generally frowned on introductions of fishes outside their native ranges. Thus, I hope that no one uses this story to support the introduction of Smallmouth or other fishes outside of their native ranges. In many cases, nonnative species wreak havoc on native ones. A clear example is the impact of the nonnative Variegate Darter on the native Candy Darter, as discussed in chapter 17, "Charismatic Candy." The introduction of Smallmouth outside of its native range has often been detrimental. In Canada, there are concerns about Smallmouth introductions into lakes with populations of Lake Trout.[23] In the United States, Smallmouth have been introduced widely west of its native Mississippi River drainage, with widespread concerns about impacts to native species, such as salmon in the Pacific Northwest.[24] Likewise, negative impacts are associated with the introduction of Smallmouth to other continents.[25]

Possibly owing to the Smallmouth's long-term residency, I have been less disconcerted about introductions of this fish than of other fishes in the central Appalachians. This complacency may coincide with my personal merging of nature and angling. As a naturalist, I am almost certain that nonnative Smallmouth integrated into the fish faunas of the central Appalachians with some level of impact on the native fishes. At the time of introductions in the 1800s, however, impacts of the nonnative Smallmouth on native fishes of the central Appalachians were not documented. In central Appalachia, the Smallmouth is a well-established fixture of the fish faunas of the Susquehanna, Delaware, Potomac, James, and New Rivers' drainages, where viable populations have existed for 100 to 165 years. As an avid angler, I have often fished for Smallmouth without giving much thought to the native or nonnative status of the species. My admiration for this species is consistent with that of Henshall, as I agree that "inch for inch and pound for pound" the Smallmouth is truly "the gamest fish that swims." As a naturalist, however, I admit to concerns about future introductions of this species, and the potential for associated negative impacts on native fish faunas.

Charismatic Candy

Animals with charisma, such as aesthetically attractive and brightly colored birds, are frequently used as flagships of nature conservation. In the central Appalachians, the common Northern Cardinal is nearly a flagship, and certainly has charisma—a species with brightly colored red males and state-bird status in Ohio, Virginia, and West Virginia. As a naturalist and ichthyologist, I admit that my admiration for cardinals has often forced my hand, trading a minnow seine for a pair of binoculars. But my favorite charismatic vertebrate of the central Appalachians is a fish. I diverge from my fellow central Appalachians who favor the Brook Trout (see chapter 12, "Trees and Trout"), although admittedly a great choice given its beautiful male coloration, popularity with anglers, and state-fish status in many regional states. My choice for central Appalachian charisma is the charming Candy Darter, a small fish with amazingly beautiful breeding males (figure 17.1). This species is native

FIGURE 17.1. A nuptial male Candy Darter from the Gauley River drainage, West Virginia, photographed on March 8, 2017.

to the upper Kanawha River system, including the New River and Gauley River drainages. After seeing the color photograph, most would probably agree that the chromatic nuptial Candy Darter is clearly charismatic, but this seldom-seen fish does not currently have flagship status. In fact, few know it exists.

Like many fishes, the Candy Darter has a history of taxonomic change. In 1932, Carl Leavitt Hubbs and Milton Bernhard Trautman described the species as *Poecilichthys osburni*: "This species is surely one of the most beautifully colored of the darters."[1] For many years, the common name of this species was Finescale Saddled Darter. This name was accurate in that the species has five or six dorsal saddles, that is, dark pigment bands that cross the dorsum (back) of the fish, and relatively small scales compared to most species of darters. But this name does not justly describe the beauty of the species. Taxonomic revision realigned the species to the genus *Etheostoma*. Subsequently, the common name was changed. In 1991, Robins et al. followed a recommendation by R. E. Jenkins and recognized the common name as "Candy Darter,"[2] reflecting the chromatic, candy-striped pigmentation pattern of the breeding males.[3]

Like many fishes, the Candy Darter is a species of conservation concern, as it has experienced geographic range reduction and dramatic population declines. At one time, the Candy Darter's geographic range was throughout the New River drainage in West Virginia, and farther upstream into Virginia within the Ridge and Valley province. It has long been extirpated from many watersheds within the New River drainage, including the Bluestone River basin, the Indian Creek watershed of the New River, and the lower section of the Greenbrier River drainage. It is difficult to point to a direct cause of its extirpation, but possibly the reduced range of this species resulted from multiple causes, such as landscape alteration, increased water temperatures, aquatic pollution, and/or competition with nonnative species.

The New River watershed of the central and southern Appalachians supports nearly 100 fish species, over half of which are nonnatives. Most of the nonnative species are attributed to bait-bucket introductions, where an angler releases live bait fishes. The New River drainage is naturally depauperate of native species, which is attributed to the geologic history of the region, particularly the presence of Kanawha Falls at Glen Lynn, West Virginia. Kanawha Falls is a barrier to the upstream movement of fishes and has isolated the fish fauna upstream of the falls from that on the downstream side. Also, because of this barrier, there is a high rate of endemism in the New

FIGURE 17.2. Some examples of endemic species of the New River drainage within the Central Appalachians, additional to the Candy Darter. Bigmouth Chub (*top left*), Kanawha Minnow (*top right*), New River Shiner (*middle left*), Kanawha Sculpin (*middle right*), Appalachia Darter (*bottom left*), and Bluestone Sculpin (*bottom right*).

River. Endemic species are those that are native within a watershed but do not occur naturally in other regions. In addition to the Candy Darter, the list of New River endemics includes Bigmouth Chub, Kanawha Minnow, New River Shiner, Bluestone Sculpin, Buckeye Creek Cave Sculpin, Kanawha Sculpin, Appalachia Darter, and Kanawha Darter (figure 17.2).[4] Additional findings will expand this list in the future. The New River endemics have small geographic ranges, making them susceptible to extinction. Populations of these New River endemics have persisted for a long time, but most have experienced range reduction. Among the New River endemics, the Candy Darter is currently a candidate for a high extinction risk.

Regardless of the reasons for the Candy Darter's decline, it is alarming to see such a beautiful fish in jeopardy. How can a charismatic animal with so much beauty be declining so rapidly, yet relatively few people seem to take notice? I suspect the Candy Darter is destined to become a flagship species, as its popularity as a species will increase with conservation awareness.

The beauty of the Candy Darter is conferred by its brilliant colors. Hubbs and Trautman described the "prevailing colors" of breeding males as "various shades of dark green and orange." Colors of the cheeks were characterized as "a rhomb of bright yellow, grading into red medially." Pigment patterns along the lateral side of the fish were reported as "greenish-blue bars" with "intervening light areas" of "deep orange." Also, the lower side under the pectoral fin was portrayed as "a brilliant horizontal blotch, grading from yellow at the edge to a brilliant orange-carmine medially; a most striking color feature." The decorative first dorsal fin was described as "largely covered by a crescent of deep greenish blue, paler on the spines than on the membranes; the basal crescent is lavender-gray, with a wedge of purplish brown on each membrane; the margin is a yellow band becoming an intense orange medially." The second dorsal fin was described as "dusky blue on the webbing, grading outward into gray within the sooty border; the gray portions of the membranes contain streaks of fire-orange; the rays themselves on their basal halves are scatteringly spotted with bright orange-red spots." The coloration of the caudal fin was characterized as similar to the second dorsal fin, but with a lighter color at the base, and with intense red spots.[5] For the remaining fins, Hubbs and Trautman denoted the coloration as follows:

> The anal fin is green-blue on membranes of basal two-thirds; the rays outward are orange yellow, producing an interrupted band; the margin of the fin is milk-white anteriorly, but becomes sooty on membranes posteriorly. The pelvic is blackish green-blue, with the edges of the rays gray, and the spine and the outer border milk-white, and with some orange near border laterally. The pectoral fin is the lightest of all and the most brilliantly spotted with fire-orange; these spots are in three or four curved grayish streaks, which separate the green-blue ground color of the rays into rather definite curved bands; the outer part of the fin is yellowish on the rays.[6]

Although the breeding colors of the male Candy Darter are accurately described by Hubbs and Trautman, I decided to delve into the digital world of computer-aided graphics. Specifically, I used my digital photographic image to name the fish's colors based on an RGB (red-green-blue) color dictionary of 1,640 names. One complication, of course, is that the digital image colors captured by my single-lens reflex camera may be biased by lighting conditions or by camera and flash settings.[7] Based on my results, the background coloration of the cheek was an orange oval of rose of Sharon, intensifying medially into Harley-Davidson orange. The pattern of barring along the lateral side included several green pigments, such as Oxley, De York, and Tom Thumb. Some pixels leaned toward blue-green aquamarine, like patina and Acapulco. The lighter bars were bordered by drover to tangerine yellow and centered with cinnabar, cardinal, and torch reds with a scattering of darker spots of guardsman red. The horizontal blotch behind the pectoral fin was bordered by yellow (saffron and golden dream) grading medially to orange and cinnabar red.

I continued the exercise with fin colors. For the first dorsal fin, the large crescent was dark spectra green grading to como green; the basal crescent was Indian tan to potter's clay brown, with a wedge of lonestar red near the bases of each membrane; the distal band was ecru white to gray nurse becoming a tawny to Harley-Davidson orange medially. The second dorsal fin had a near-basal stripe of dark blue Dianne to firefly green on membranes, and light blue-green (Gulf Stream) on the rays bordered distally with spots of sienna brown to orange (grenadier, rock spray, and red stage). The middle section of the second dorsal fin had a series of brown colors (paco, space shuttle, mondo, and makara) with membrane strips of fire orange to rose of Sharon orange, and a distal border of yellow (dark goldenrod and gold) or orange (tawny and mango tango). The penultimate border strip of the second dorsal fin was Dutch white with a gray terminal border (eclipse, gravel, and tuatara) to green (corduroy, Xanadu, and shadow). As observed by Hubbs and Trautman, the coloration of the caudal fin is characteristic of the second dorsal fin, but with lighter colors and intense spots of fire brick red and rose of Sharon orange. For the basal half of the anal fin, membranes are cadet blue to Gulf Stream green and rays are hippie to fern green. Red, orange, and yellow markings on the anal fin are present as cinnabar red, Chilean fire, and Harley-Davidson orange, and gold, turbo, and energy yellow. The posterior border of the anal fin is sirocco to patina green with a narrow strip of soapstone gray, and the anterior border is white to soapstone gray. The pelvic fin has rays of green (clover, Tom Thumb, and chalet) with membranes of cadet blue to patina and ming green; near-border strips of bronze, festival yellow, and golden sand yellow; and a border of quarter pearl lusta to white. The pectoral fin has spots of Trinidad orange to medium carmine and a brilliant alizarin red, with outer rays of straw to marzipan yellow.

The color brilliance of male Candy Darters is mind-boggling, likely explained in part by sexual selection. In 1859, Charles Darwin proposed sexual selection,[8] with further refinement in his 1871 book *The Descent of Man, and Selection in Relation to Sex:* "We are, however, here concerned only with that kind of selection, which I have called sexual selection. This depends on the advantage which certain individuals have over other individuals of the same sex and species, in exclusive relation to reproduction."[9]

In 1994, Malte Andersson provided a consistent but more concise wording in his book *Sexual Selection:* "Sexual selection arises from differences in reproductive success caused by competition for access to mates."[10] In reference to the Candy Darter, mating involves choosy females and competitive males, where the female has the power of mate choice. Larger and more attractive males are more likely to be chosen by females, thus outcompeting other males for mates. This Darwinian view is based on the attractiveness of beautiful males, but Darwin's contemporary Alfred Russel Wallace proposed that more attractive males are also likely to be healthier males, thus mate selection may be based more on male vigor than on beauty.[11] Regardless of the reasoning behind female mate choice, I suspect that sexual selection is at play in Candy Darter mating success. It would be reasonable to assume that bright colors associated with sexual selection would come with increased predation risk. However, high predation rates have likely not caused the range reduction and population declines of the Candy Darter, as it lives in stream segments of rocky riffles with fast-flowing waters and low predator densities.

A recent conservation concern for the Candy Darter is that it is hybridizing with the closely related, nonnative Variegate Darter. Angler bait-bucket releases have been attributed to the introduction of the Variegate Darter into the New River drainage. Interestingly, Jared Potter Kirtland, in his 1840 species description of the Variegate Darter, stated: "It is frequently taken by fishermen for bait, and preferred to the common minnows."[12] As an avid angler, I have admiration for all of my rod-and-reel-toting colleagues of the central Appalachians. Anglers represent a diverse group of individuals, many of whom are knowledgeable about the natural history of fishes. Most of my angling friends have a strong conservation ethic, where importance is placed on protecting the resource, emphasizing clean water and good habitat. As in all groups, though, some individuals have bad habits. One habit, rarely recognized as "bad" by anglers, is the moving of fishes between drainages. This scene is played out repeatedly in the central Appalachians, where an angler seines or traps live fishes in one watershed, bucket-transports these fishes for use as fish bait in an adjacent or distant drainage, then releases the leftover bait at the end of the fishing trip. In some cases, several fish of the same species released from a bait bucket function as a founder population, where survival and successful reproduction lead to an established and then expanding nonnative population. If successful, a nonnative fish first becomes established as a small localized population, often followed by rapid range expansion. Nonnative species, with high densities and expanding distribution ranges, commonly compete with native species for habitat and food.

In some cases, following the introduction event, nonnative species hybridize with native species. Hybridization is particularly common in fishes when the two forms are close relatives. It occurs when the individuals of two species interbreed and produce offspring (called an F1) with a genetic composition derived from both parents. If hybrid individuals are not sterile, then genetic transfer occurs with subsequent mating between the parental and hybrid individuals. This transfer of genes is called introgression. Over time, with continued introgression, genes of the native species can be swamped, resulting in the extinction of the native species.

The exact date of the introduction of the Variegate Darter upstream of Kanawha Falls is not known. The first record of the species in the New River was documented in 1982 at Sandstone Falls by the ichthyologist Daniel A. Cincotta.[13] From 1991 to 1996, five specimens were collected from the New River at Sandstone Falls, and these individuals shared characters of Candy and Variegate darters. In 1993, a Variegate Darter was collected

FIGURE 17.3. A male Variegate Darter photographed on April 19, 2017, from Twentymile Creek, a tributary of the Gauley River.

from Anthony Creek (a tributary of the Greenbrier River). In 1995 and 1999, Variegate and Candy Darters were present in Anthony Creek. In 1997, Variegate Darters and hybrids were collected from the New River at Stone Cliff Bridge.[14] In 1999, a large population of Variegate Darters was found in the Greenbrier River at Ronceverte. In 2001, Variegate Darters were widespread within the New River gorge.[15] During 2004 and 2018, Candy Darters, Variegate Darters, and hybrids were collected from the New and Gauley Rivers' drainages (figures 17.3, 17.4), and fin clips from those specimens were used for genetic study.[16]

Introgressive hybridization, resulting from anglers' bait-bucket introductions, may bring the Candy Darter to the brink of extinction, but there is still hope. While I wrote this chapter, several conservation efforts for this fish have come to fruition. Biologists with the US Fish and Wildlife Service and West Virginia Division of Natural Resources are currently collaborating on captive propagation efforts to restock hatchery-reared genetically pure Candy Darters into areas of previous extirpation. Additionally, in 2018, critical habitat and endangered species status were designated for the Candy Darter by the US Fish and Wildlife Service.[17] The Candy Darter, with its current conservation status, will likely achieve wider recognition as a charismatic species. Recently, I have seen images of the Candy Darter on conservation-oriented stickers, T-shirts, key chains, a beer label in Fayetteville, West Virginia, and as a print on the backstrap of a pair of Chacos.

FIGURE 17.4. A male F1 hybrid (Candy Darter × Variegate Darter) photographed on March 7, 2017, from Deer Creek, a tributary of the Greenbrier River.

The Candy Darter's image on a US Postal Service stamp will soon be distributed around the world. Perhaps the species will achieve "flagship" status, but hopefully not posthumously. I cannot predict if the Candy Darter will persist or perish, but I remain optimistic for its persistence and ultimate recovery.

Like a Rolling Stone

Muddy Waters, the renowned blues musician, sang about catfish in his 1950 single "Rollin' Stone," a song that influenced the names of a band, a magazine, and one of my favorite Bob Dylan songs. Like a song, nature influences people in various ways. I, for example, often find nature in songs. The song "Rollin' Stone" returns my thoughts to the feeding behavior of one of my favorite fishes, the Logperch. Using an unusual tactic for finding food, the predatory Logperch scrounges for its next meal by using its conical snout to flip small flat rocks and roll small round stones. By moving stones, the Logperch exposes its primary prey, small insect larvae such as mayflies, stoneflies, and caddisflies.

The renowned naturalist Louis Agassiz is often quoted with the adage "Study nature, not books." I have taken Agassiz's advice by snorkeling in central Appalachian streams, often with an interest in the ecology and behavior of darters, a diverse group of colorful fishes in the family Percidae, which includes the Logperch. During summer, most darters of the central Appalachians live in riffles, the fast-flowing, gurgling, rocky, shallow sections of streams. Riffles are desirable to darters because of the high levels of oxygen in the water, the low numbers of large predators in shallow water, and the protective cover of rocks.

Although you can stand on the stream bank and see darters in riffles, full appreciation of these small and colorful fishes requires an underwater view with a dive mask and a snorkel. The fast-flowing water of a riffle feels cool, even on a warm summer day, so I often snorkel while wearing a wetsuit made of thin neoprene fabric. The wetsuit also protects elbows and knees from the abrasive rocks of the stream bottom. Snorkeling in shallow riffles involves crawling on the stream bottom with your head or face submerged. The water is too shallow for swimming, often less than two feet (61 cm) deep. The musical gurgling sounds of a riffle, as heard from the stream bank, are magnified underwater. Not surprisingly, both above- and underwater sounds of riffles are marketed as relaxing sounds of nature for meditation. I agree that riffle sounds are soothing, but my satisfaction during a day of snorkeling comes from seeing the dazzling display of darters—darters peering out from under rocks, swimming between and over rocks, feeding on small insects, and occasionally darting under rocks to avoid a predator.

Some species of darters remain stationary while allowing water currents to deliver food items, a feeding strategy referred to as "lie-in-wait." Other fishes actively forage. Based on my underwater observations, Logperch occasionally congregate in schools, but are often solitary feeders on or near the stream bottom. The Logperch often forages actively by rolling stones. One tradeoff for small fishes that actively forage, however, is that they cannot always lie in wait under or near protective cover, but must sometimes be out and about, exposed to larger predators. One large predator, the Smallmouth Bass, commonly consumes small fish, and Logperch is listed on the menu. Thus, in the presence of Smallmouth Bass, the Logperch is not able to focus full attention on finding food. So it seems that Logperch must search simultaneously for both life-sustaining food and deadly predators. Interesting tradeoffs in nature often lead to simple but interesting questions: how do small fishes actively forage while avoiding larger active foragers?

Although stone rolling is fascinating, my fondness for Logperch also results from its physical appearance. The Logperch pops with striking patterns of pigmentation. When talking about pigment markings on animals, you may first think of extremes, like the patterns of peacock feathers, a leopard print, or zebra stripes. Fishes, which are often submerged from public view, can also display dazzling body markings. From a lateral view, the Logperch is zebra-barred with around 20 dark vertical narrow bars of pigment (figure 18.1). Like many fishes, Logperch have dark dorsums (backs) and light ventrums (undersides). From a dorsal view, the zebralike bars join to form saddles—bands that cross the dorsum. Four saddle bands are wider than the rest. Also, a dark round spot of pigment is present at the base of the tail (caudal) fin. Similar to lateral body pigment, both dorsal and caudal fins are barred. The zebralike stripes, dorsal saddles, and dark basicaudal spot combine to create a conspicuously attractive species.

At a maximum adult size of about seven inches (18 cm), the Logperch is a large darter, but is small relative to several other fishes within the stream community, including large fish-eating predators like adult Smallmouth

FIGURE 18.1. Photograph of a 4.9-inch (124-mm) Logperch from a Central Appalachian River.

Bass. Rock flipping and stone rolling require the Logperch to be in the open, exposed to these larger predators. Intuitively, you may wonder whether bands, saddles, and spots would increase visibility of the Logperch, thus increasing its vulnerability to predators. To rephrase an earlier question: how does a conspicuously pigmented small fish simultaneously roll stones and avoid predation?

Appearances can be deceiving. The conspicuous pigment patterns that I see on the Logperch are not seen the same way by underwater predators. Instead, some pigment patterns camouflage and conceal Logperch. The benefits of camouflage are obvious—nondetection by predators. If individual Logperch survive to adulthood (typically at age two), then they can reproduce and contribute to the next generation. In this way, a population of Logperch persists over time.

Although the benefit of camouflage may be obvious, an explanation of how it works is not. You could assume that an animal's pigment pattern works by mimicking or matching the surrounding habitat, an explanation known as "background matching." In nature, habitats and backgrounds are often not uniform, but rather represent a complex of many shapes, shades, and shadows. If an animal's survival depends on background matching, then it must live in or on habitat with a specific background. This limits the animal's movement, because if it moves away from the background then it becomes vulnerable to predators. Often animals need to move to find food. A camouflaged animal relying solely on background matching must forage within a background-specific area, or chance predation while foraging elsewhere.

Previously, some students of camouflage have leaned errantly on a single explanation. Interestingly, this occurred, in part, in the later work of Abbott Handerson Thayer, one of America's great renaissance men, known as a painter of women, angels, and landscapes; a teacher; an amateur naturalist and ornithologist (the study of birds); a conservationist; and the father of camouflage.[1] In 1903, Thayer offered an interesting statement regarding animal markings: "Nature has evolved actual art on the bodies of animals, and only an artist can read it."[2] Thayer was often painted by his peers as an eccentric and an extremist. In the book *Concealing-Coloration in the Animal Kingdom,* coauthored with his son in 1909, Thayer stated that "all markings and patterns whatsoever are, under ordinary outdoor conditions, unfavorable to the conspicuousness of the thing that wears them."[3] Based on this statement, former president Teddy Roosevelt criticized Thayer in 1911,[4] in part for promoting background matching to the extreme.[5] Some of Thayer's views were quite controversial, including the blue feathers of the blue jay matching the blue background of snow and the pink feathers of flamingos blending with the red sunset of the African sky.[6]

Despite Thayer's extreme views, he was well aware of alternative explanations of camouflage. In addition to background matching, Thayer pioneered two other currently supported views on animal concealment: countershading and disruptive coloration. Countershaded animals are camouflaged by the effect of sunlight and shadows on dark backs and light bellies.[7] The dark back is lightened by the sunlight, and the light belly is darkened by the shadow. This effectively results in a uniform color, removing body dimensionality and producing the perception of flatness. In his 1896 publication "The Law Which Underlies Protective Coloration," Thayer depicted countershading as "a beautiful law of nature," "the law of gradation in the coloring of animals," and further stated, "Animals are painted by nature, darkest on those parts which tend to be most lighted by the sky's light, and *vice versa*."[8] Countershading concealment is sometimes called Thayer's law.[9]

Disruptive coloration, a term Thayer introduced as "ruptive," occurs when bold markings disrupt or break up an animal's body outline.[10] A disruptively colored animal has markings of high contrast along the perimeter of its body.[11] Disruptive coloration conceals an animal's body shape. Unlike background matching, disruptive coloration is not background specific. A disruptively colored animal maintains its camouflage while moving and foraging across habitats.

Logperch likely achieve concealment by combining background matching, countershading, and disruptive coloration. From a distance, a predator's view from the topside merges the Logperch's dark dorsum with the dark shadows of the streambed, an example of background matching. From a closer view, dorsal saddles hide fishes in benthic habitats because predators see the lighter dorsum areas between dark saddles as "rocks" and the darker saddles as "shadows between rocks." In streams, predatory fishes do not always spot benthic prey from above, but may approach from the side. From the side, the dark dorsum and light ventrum of the Logperch lead to concealment by countershading. A fish is countershaded when downwelling light illuminates the dark dorsum and its upper half shades the light ventrum. The result is a fish with near invisibility from a lateral view. Further, concealment is enhanced by the disruptive coloration of the zebralike stripes and dorsal saddles, a disruption of body outline. From a side view, the vertical bars break up the Logperch's body outline. Interestingly, a fish can often control the expression of its pigment. In Logperch, the lateral vertical bars are darkest during daylight, but dorsal saddles become more prominent during nighttime, when the Logperch rests on the bottom.

The Logperch's camouflage is also influenced by how often the bars repeat per unit of distance, an effect referred to as spatial frequency. Here we dive into physics, mathematics, and optical illusions. For a pattern of vertical bars, spatial frequency increases with decreases in both the width of the bars and the width of the spaces between the bars (figure 18.2). Spatial frequency can be measured as cycles per unit of distance.[12] A pattern of bars at a

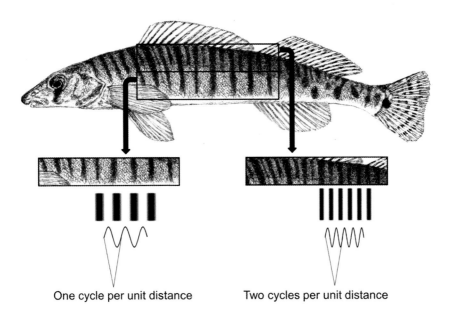

One cycle per unit distance   Two cycles per unit distance

FIGURE 18.2. Spatial frequency is a measure of how often vertical stripes repeat per unit of distance. In this example, spatial frequency increases with decreases in both the width of the bars and the width of the spaces between the bars. Spatial frequency can be measured as cycles per unit of distance.

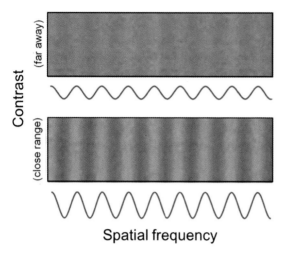

FIGURE 18.3. Contrast, influenced by distance or light, affects the spatial frequency of vertical stripes. An object with vertical stripes has low contrast (appears more uniform) when viewed from a far distance or during darker conditions.

single spatial frequency is perceived differently at differing distances. When viewed from afar, a low spatial frequency is perceived as a smoothly varying gray scale with low contrast. At close range, the same spatial frequency is perceived with higher contrast (figure 18.3).[13] The spatial frequency of the Logperch's vertical bars is especially interesting from a predator's view. Prominent vertical bars extend from the dorsum to the lower side of the Logperch, but weaker bars (dorsum to upper side) occur between the prominent bars (figure 18.1). A predator's perception of spatial frequencies of this two-bar pattern differs depending on time of day, distance between predator and prey, and positioning (benthic vs. pelagic). All bets should be placed on the prey. The Logperch's predation risk is likely reduced by the predator's perceived spatial frequencies of the two-bar pattern.

Camouflaged prey are not invulnerable to predators, although stealthy predators often stalk stone-rolling Logperch without success. In one scenario, a predator spots a Logperch and starts pursuit. At this point, it would appear that the Logperch's survival depends solely on both parties' swimming speeds. Interestingly, vertical bars may play a role during escape. A pattern of vertical stripes may reduce capture probability of moving prey by altering a predator's perception of speed.[14] Owing to the spatial frequency of vertical stripes, the swimming speed of the Logperch may be perceived by the predator as slower than actual. A common example of this optical illusion, known as the wagon-wheel effect, occurs when a spoked wheel in motion appears to either rotate backwards, stand still, or move more slowly than its actual rotation speed.[15] In another

optical illusion, the direction or trajectory of a moving object with vertical bars can appear to move at an angle that differs from the true trajectory.[16] Thus, optical illusions caused by the movement of vertical stripes
may prevent predators from accurately judging the speed and trajectory
of the escaping Logperch. In this way, a pigment pattern may attract and
confuse a predator.

By optical illusion, stripes also alter a predator's perception of prey
size. In humans, vertically striped clothing is commonly thought to create an illusion of thinness, but research discredits this as a myth.[17] This
can be illustrated by the Helmholtz square illusion, named after Hermann
von Helmholtz, a German physician and physicist who studied visual
perception. The illusion occurs when viewing two adjacent equally sized
squares, one with vertical stripes and the other with horizontal stripes.
The square with vertical stripes appears to be wider, and the square with
horizontal stripes appears to be taller and narrower (figure 18.4). Possibly
a Logperch may appear longer and larger to a predator because of its vertical stripes.[18] A prey's perceived size could save its life because fishes are
generally gape-limited. Prey larger than the predator's mouth gape may
be ignored. Intuitively, based on the Helmholtz square illusion, a predator
may misjudge the size of a striped Logperch.

The illusion of size associated with vertical stripes may also alter the
predator's perceived distance to prey, which plays a part in determining when

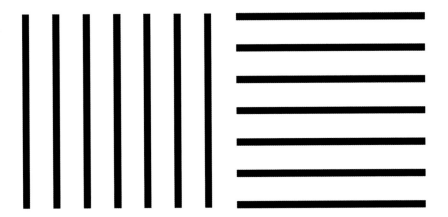

FIGURE 18.4. The Helmholtz square illusion occurs when viewing two adjacent equally
sized squares, one with vertical stripes and the other with horizontal stripes. The square
with vertical stripes appears to be wider, and the square with horizontal stripes appears to
be taller and narrower.

FIGURE 18.5. In an illusion of depth perception, an object with vertical stripes appears closer than actual when placed in front of a complex background, such as the rocky habitat of a stream bottom.

the predator begins pursuit. This illusion of depth perception is illustrated by placing an object with vertical stripes in front of a complex background (figure 18.5).[19] The object with vertical stripes appears closer than the background. If predator-to-prey distance is underestimated, then the predator may start pursuit too soon, allowing the prey time to react and escape.

In another example of deception, a conspicuous dark round spot is present posteriorly on some fishes. The Logperch has a prominent spot at the base of its tail that may masquerade as an eye, misleading a predator to strike at this "eyespot" instead of the more vulnerable head and torso. Researchers call this the deflection hypothesis.[20] Another explanation is that "eyespots" do not mimic eyes, but are just conspicuous spots that attract a predator's attention.[21] With either explanation, small fishes attacked at the caudal end are more likely to survive than those attacked at or near the head. As with most fishes, the eyespot is generally more pronounced in young Logperch, a time when predation risk is highest. In the central Appalachians, many fishes have basicaudal eyespots, including some common fishes like Bluntnose Minnow, Creek Chub, Spotted Bass, and Fantail Darter. In many fishes, including the Logperch, the actual eye is often obscured by a subbar of pigment, known as a subocular or infraorbital bar (figure 18.1). Some fishes also have pre- and postorbital pigment bars that mask the eye.

Cryptic and conspicuous body pigment patterns are not unique to the Logperch, but in fact are common among the globe's estimated 35,000 fish species. I saw similar patterns of crypsis in fishes of the Great Barrier Reef, off eastern Australia. While snorkeling there, I witnessed countershading, vertical barring, dorsal saddles, and eyespot mimicry. Like our Logperch, zebra-barred reef fishes exploit the illusory physics of vertical bars to escape from predators. Despite the grand diversity of reef fishes, however, few fishes rival the suite of cryptic and illusory pigments of our Logperch.

The Logperch is perhaps the prince of predator avoidance, with all previously mentioned pigments: countershading, dorsal saddles, vertical barring, and basicaudal eyespot. From all angles of attack, pigment patterns of the Logperch possibly offset some of its predation concerns during pelagic positioning and benthic stone rolling. Camouflaged by dorsal saddles, the Logperch reduces its risk to overhead predators. The Logperch as a barred, obliterative countershader reduces lateral predation risks. With its head down in rock-flipping position and caudal fin upward, the Logperch's basicaudal eyespot is also a predation-risk reducer. Further, the optical illusion of the vertical bars may confuse predators about the escaping Logperch's speed, trajectory, and body size.

Predator avoidance, in part, possibly explains the selective advantages of dorsal saddles, eyespots, and zebra stripes. I speculate that camouflage, eye mimicry, and the illusory physics of vertical bars all contribute to the Logperch's survival and its foraging success while stone rolling. To my knowledge, however, none of the explanations above have been experimentally tested with Logperch. I hope that someone will take these research questions a step further and test these explanations. Possibly none, some, or all are true, but as of now it is "like a complete unknown." I may or may not get around to these further investigations. In the meantime, I will be content to listen to Muddy Waters, Bob Dylan, and the Rolling Stones, and continue snorkeling with further observational studies and speculations about the "nature" of stone-rolling Logperch.

Tit-for-Tat

I often find myself immersed in nature, particularly when snorkeling in rivers of the central Appalachians. River snorkeling provides a unique opportunity to observe fish behavior. Once, as a graduate student, I observed an unusual behavior, one that I did not fully understand until later in life. While immersed in the river, I was closely watching the swimming behavior of a six-inch (15.2-cm) Smallmouth Bass (figure 19.1). This species is widely distributed across the central Appalachians, occurring commonly in medium to large rivers (see chapter 16, "Gamest Fish That Swims"). The Smallmouth Bass, with slight and seemingly effortless fin movements, turned to look toward the stream margin. I followed its gaze, and in the crystal-clear water we both began watching a foraging shoal of seven Dusky Darters (figure 19.1).

My interest turned to darter behavior, because all of my previous Dusky Darter encounters were of single individuals, never shoals. A shoal of fishes forms when individuals aggregate and orient in different directions, sometimes foraging together as a group. The shoaling darters were about three inches (7.6 cm) in length. Small fishes such as darters are often focused on their surroundings for the presence of predators, even while foraging. Thus, fishes in a shoal must balance two conflicting demands: maximizing foraging success while minimizing the risk of predation.[1]

In this triangular encounter, the Smallmouth Bass was about three feet (91 cm) from my facemask, and the Dusky Darters were about four feet (122 cm) from me and the Smallmouth Bass. The darters closely watched the bass, while I waited with expectations of observing the bass' predatory behavior. The bass did not pursue the darters. Instead, two darters left the group and began to approach the bass. My initial reaction was to question this seemingly ill-advised behavior. Why would two small darters leave the safety of

FIGURE 19.1. A Smallmouth Bass (*top*) about five inches (12.7 cm) in length, and a Dusky Darter (*bottom*) about 2.5 inches (6.4 cm) in length.

the shoal to approach a potentially life-threatening predator? The two darters were nearly side by side in their approach, with one slightly lagging behind the other. The lead darter moved a few inches, then the trailing darter moved a few inches. In this copycat scenario, the two darters approached the bass with an iteration of short movements, stopping at about one foot away from the bass. The bass still did not pursue the two darters. Then the lead darter slowly turned and returned to the shoal, closely followed by the other darter. Apparently the bass was sated; its motivational state was to observe and not to attack. The shoaling darters foraged on, as if no threat were present. Possibly the returning duo had delivered a message to their shoalmates, that the current threat level was low. Otherwise their decision to approach the bass was a mystery.

Sometime later I stumbled onto a published paper that discussed an interesting behavior associated with shoaling fishes. The title of this paper was "Dicing with Death: Predator Inspection Behaviour in Minnow Shoals."[2] I knew that fishes often form shoals, a common behavior that reduces the probability of predation of members of the group. What I did not know was that when shoaling fish are being surveilled by a predator, one or more individuals will sometimes leave the safety of the shoal and swim toward the predator to assess the threat level. This interesting behavior, called "predator inspection," likely explained the Dusky Darter–Smallmouth Bass interaction. With further reading, I learned that this behavior was well documented for many small-bodied, shoaling fishes,[3] where inspection of potential predators can occur by one, two, or occasionally three or more individuals.[4]

Predator inspection behavior has a mathematical interpretation. Consider the contributions of John Forbes Nash Jr., a mathematical genius and native of central Appalachia. Nash was born and raised in Bluefield, West Virginia. While at Princeton University, he developed an elegant theorem, known as the Nash equilibrium, now included within a large field of research called "game theory." Nash's work became more widely known with the 2001 biographical drama film *A Beautiful Mind,* where he was played by the actor Russell Crowe. Nash's theorem has had wide application, leading to his award of the 1994 Nobel Memorial Prize in Economic Sciences. His equilibrium theorem was applied in political science by Robert Axelrod,[5] and in evolutionary theory and animal behavior, including the concept of an evolutionary stable strategy, by John Maynard Smith.[6]

A particularly interesting application of game theory is known as the Prisoner's Dilemma. Hofbauer and Sigmund summarized the game as follows:

> The Prisoner's Dilemma (or PD) game is a game between two players, each having a choice between two options: to cooperate (to play **C**) or to defect (to play **D**). The payoff matrix is given by
> If both players cooperate, they receive as payoff R (the reward), which is assumed to be larger than the payoff P (the punishment) obtained if they both defect. But if one player plays **D** while the other player plays **C**, then the defector receives a payoff T (the temptation) which is even higher than R, whereas the cooperator receives only the sucker's payoff S which is lower than P.[7]

The Prisoner's Dilemma game can be confusing at first, but consider our two Dusky Darters as players of the game. Also, consider a summary of the previous paragraph, where the payoff values are ranked $T > R > P > S$. If the two Dusky Darters (predator inspectors) are in a Prisoner's Dilemma, then their options are to either cooperate or defect. If one darter makes the first move toward the predator, then the other may cooperate by making the same choice, resulting in a payoff of R (both are rewarded for mutual cooperation). If both cooperate on each move, then the two darters approach the predator, assess the threat, and return to share the information with their shoalmates. Alternatively, when one darter moves toward the predator, the other darter may choose to defect (not make a move toward the predator). A darter that selfishly defects is rewarded because its predation risk is reduced by abandoning the predator inspection. The payoff for the darter that defects can be thought of in terms of a benefit on fitness (survival). In this case, the defecting darter receives the highest payoff of T (temptation to defect). Because $T > R$, a darter's best move is to defect. If one darter defects, then the other darter either receives the lowest payoff (S) or chooses to defect, resulting in the payoff of P for both individuals (punishment for mutual defection). When one darter defects, then the other darter's best choice is to also defect, because $P > S$. Interestingly, it seems that a darter's best strategy would be to defect, regardless of what the other darter decides to do. The "dilemma" is that if both darters defect, then both get a payoff of P, which is less than the payoff (R) for mutual cooperation.

If the best strategy is to defect, then why did the two Dusky Darters cooperate? If two individuals interact only one time in a Prisoner's Dilemma game, then the best move is always to defect. But in an iterative Prisoner's Dilemma, with multiple successive interactions, cooperation is often the best strategy.[8] In this case, a common and successful cooperative strategy of a series of games of Prisoner's Dilemma is called Tit-for-Tat.[9] As noted by Axelrod and Hamilton, "This strategy is simply one of cooperating on the first move and then doing whatever the other player did on the preceding move."[10]

The Tit-for-Tat strategy to the iterative Prisoner's Dilemma game offers a fascinating interpretation of predator inspection. Cooperation is a winning strategy. Predator inspectors gain and transfer information on the threat of predation, presumably reducing risk for other group members. If both darters defect on the predator inspection, then no information about the predator is transferred to the shoalmates. Without this information, shoalmates could lose valuable foraging time owing to time spent suspiciously watching a nonthreatening predator. Additionally, if both darters defect when the threat is real, then shoalmates may focus on foraging and be at a higher risk of predation.

It is reasonable to assume that predator inspection is related to improved fitness (survival). As noted by Godin and Crossman, "For prey to approach their predators seems paradoxical, since such behaviour presumably renders the prey more conspicuous to the predator and more vulnerable to attack by being in closer proximity to the latter (Dugatkin 1992; Dugatkin and Godin 1992; Pitcher 1992). For such a behaviour to evolve and be maintained in a population, its average fitness benefits must exceed or equal its associated average fitness costs, and it may also have to provide a greater benefit-cost ratio than that of an alternative tactic (cf. Dugatkin and Godin 1992)."[11]

Instead of the Tit-for-Tat strategy of two predator inspectors, would it be simpler for a single fish to inspect a predator? First, although two inspectors may get the same information about a predator, the predation risk per individual with two fish is half that of a single fish.[12] Thus, the "safety in numbers" advantage of two fish is safer than the single-fish strategy. Second, studies support that two fish will often approach closer to a predator than a single fish.[13] A closer approach may yield more accurate information about the predation threat.

Information gained from a predator inspection likely includes the predator's identity, the proximity of the predator to the shoal, and the behavior of the predator.[14] The shape and orientation of a predator's body can make it difficult to detect from a distance, but by predator inspection a fish can get a closer view of shape and markings that identify a predator. This identity check is important because the inspecting fish may discover that the large fish looming in the distance is benign. If predator status is verified, then the inspecting fish may be able to assess the predator's behavior and its attack motivation.[15] Additionally, according to Pitcher, Green, and Magurran, "It is possible that inspection behaviour informs the predator that it has been spotted and therefore helps to inhibit an attack, since the predator is aware that the element of surprise has been lost."[16]

Consider that some of the social interactions of fish in nature may transfer to our own experiences. As stated by Axelrod and Hamilton, "Many of the benefits sought by living things are disproportionally available to cooperating groups. While there are considerable differences in what is meant by the terms 'benefits' and 'sought,' this statement, insofar as it is true, lays down a fundamental basis for all social life. The problem is that while an individual can benefit from mutual cooperation, each one can also do even better by exploiting the cooperative efforts of others. Over a period of time, the same individuals may interact again, allowing for complex patterns of strategic interactions."[17]

In our one-on-one relationships with friends and colleagues, these repeated "complex patterns of strategic interactions" may consistently yield higher payoffs if we cooperate more often. Just something to consider.

Discovering Diamonds

Part of being a naturalist is that people are interested in hearing about your scientific discoveries. During an interview, I once stated: "If there is one fun and exciting thing about science, then it is the discovery of new things." Perhaps the words "thing" and "things" were not the most scientifically eloquent, but the sequence of "science" and "discovery" secured my emphasis. The interview was about the discovery of the Diamond Darter (*Crystallaria cincotta*) from the Ohio River drainage (figure 20.1), a species new to science in 2008.[1]

The Diamond Darter is a small streamlined fish, with adults growing to lengths of nearly four inches (10.2 cm). Distinct markings on the body

FIGURE 20.1. A male Diamond Darter about 3.5 inches (89 mm) in length from the Elk River, West Virginia.

(pigment patterns) are useful for identifying this fish. Four distinct saddles of darker pigmentation are present on the upper sides, crossing over the back. The lower sides of the head and body are silvery white. The eyes are positioned high on the head, and a dark blotch of pigment occurs in front of the eye. A series of 12–14 oblong blotches of pigment is present along the midlateral side. Other interesting characters include a sickle-shaped posterior margin of the pelvic fin, a relatively large mouth, and the absence of scales on certain parts of the body. The breast and belly are without scales, and the side of the head has just a few rows of scales.

When thinking about the discoveries of new fish species from the Ohio River basin, I am reminded of the nineteenth-century naturalist Constantine Samuel Rafinesque (1783–1840), a discoverer of new fishes, some of which currently occur in central Appalachian streams.[2] Rafinesque's descriptions of new species from the Ohio River drainage were published during a four-year period (1817–20). In his 1820 publication titled *Ichthyologia Ohiensis: Natural History of the Fishes Inhabiting the River Ohio and Its Tributary Streams,* Rafinesque wrote:

> Nobody has ever paid any correct attention to the fishes of this beautiful river, nor indeed of the whole immense basin, which empties its water into the Mississippi, and hardly twelve species of them had ever been properly named and described, when in 1818 and 1819, I undertook the labour of collecting, observing, describing, and delineating those of the Ohio. I succeeded the first year in ascertaining nearly eighty species among them, and this year I added about twenty more, making altogether about one hundred species of fish, whereof nine tenths are new and undescribed.[3]

In 1886, David Starr Jordan referred to Rafinesque as a "monomaniac on the subject of new species."[4] Rafinesque's reputation as a naturalist has been questioned since his own day, although others have highly valued his contributions to science.[5] Despite Rafinesque's controversial character, I would argue that his ichthyological legacy is very much alive in the central Appalachians. His feverish focus on the discovery of new fishes, as well as other discoveries of new animals and plants, was influential toward understanding the flora and fauna of the central Appalachians.

Today Rafinesque is credited with the discovery of 36 fish species from the Ohio River drainage (appendix 3),[6] but many of his species are no longer recognized. In some cases, a cluster of similar species described by

Rafinesque was later realized to represent a single species. In other cases, he described species based on illustrations or verbal information from other people. Perhaps part of Rafinesque's discredit comes from new species descriptions based on fictitious illustrations—as in a revenge prank played on him by John James Audubon,[7] the celebrated American ornithologist, naturalist, and painter of birds.[8]

In 1818, Rafinesque traveled on foot and boat to visit Audubon, a long journey from Philadelphia to Henderson, Kentucky. In 1832, Audubon described Rafinesque as the "eccentric naturalist," describing his appearance in detail:

> A long loose coat of yellow nankeen, much the worse for the many rubs it had got in its time, and stained all over with the juice of plants, hung loosely about him like a sac. A waistcoat of the same, with enormous pockets, and buttoned up to his chin, reached below over a pair of tight pantaloons, the lower parts of which were buttoned down to the ankles. His beard was as long as I have known my own to be during some of my peregrinations, and his lank black hair hung loosely over his shoulders. His forehead was so broad and prominent that any tyro in phrenology would instantly have pronounced it the residence of a mind of strong powers. His words impressed an assurance of rigid truth, and as he directed the conversation to the study of the natural sciences, I listened to him with as much delight as Telemachus could have listened to Mentor.[9]

It seems that Rafinesque did not dwell on appearance, which seems reasonable for a man chasing discoveries. Rafinesque's passion for discovering and describing new species was revealed in his reaction to a plant along the riverbank. Audubon wrote: "He plucked the plants one after another, danced, hugged me in his arms, and exultingly told me that he had got not merely a new species, but a new genus. When we returned home, the naturalist opened the bundle which he had brought on his back, and took out a journal rendered water-proof by means of a leather case, together with a small parcel of linen, examined the new plant, and wrote its description."[10]

While Audubon housed and hosted Rafinesque, he further realized Rafinesque's ravenous desire to discover new species:

> We had all retired to rest. Every person I imagined was in deep slumber save myself, when of a sudden I heard a great uproar in the

naturalist's room. I got up, reached the place in a few moments, and opened the door, when to my astonishment, I saw my guest running about the room naked, holding the handle of my favorite violin, the body of which he had battered to pieces against the walls in attempting to kill the bats which had entered by the open window, probably attracted by the insects flying around his candle. I stood amazed, but he continued jumping and running round and round, until he was fairly exhausted, when he begged me to procure one of the animals for him, as he felt convinced they belonged to "a new species." Although I was convinced of the contrary, I took up the bow of my demolished Cremona, and administering a smart tap to each of the bats as it came up, soon got specimens enough.[11]

Subsequently, Audubon performed an unnatural and unprecedented practical joke. Perhaps in revenge for his crushed Cremona, he illustrated at least 11 fictitious fishes, presenting the drawings along with descriptive stories to Rafinesque.[12] Trusting Audubon, Rafinesque presumed the illustrations were accurate and described the faux fishes as new to science. Rafinesque clearly made a mistake in describing these species without first seeing specimens. The hoax, however, reflected as poorly on Audubon's character as it did on Rafinesque's reputation as a naturalist. David Starr Jordan wrote: "In regard to the descriptions of fishes made by Rafinesque from 'drawings by Mr. Audubon,' I am informed by Dr. Kirtland, on the excellent authority of Dr. Bachman, that several of the monsters described by Rafinesque (such as *Aplocentrus, Pogostoma, Eurystomus,* etc.) were drawn by Audubon with a view to a practical joke on the too credulous ichthyologist. That being the case, it is but justice to Rafinesque's memory to let those names drop from our systematic lists without prejudice to him."[13]

Scientists have subsequently made many new discoveries about the species Rafinesque described. Eccentric as he was, Rafinesque was similar in some ways to other naturalists, as all are interested in discovery. But with new findings, there are always more questions to ask. Finding a new species always leads to new questions, often answered with new discoveries on life history, habitat use, geographic distribution, and ecology. Similarly, with the discovery of the Diamond Darter, I have worked with graduate students and colleagues on many questions related to habitat use, distribution, and reproductive ecology. What new information have we found about this species?

First, the story of the Diamond Darter and associated discoveries is almost as long-winded as the writings of Audubon. On November 13, 1980, Mike Hoeft of the West Virginia Division of Natural Resources (WVDNR) discovered an unusual fish during a survey of the Elk River near Mink Shoals, West Virginia. Daniel A. Cincotta, a WVDNR fish expert, realized that this was a rare find—a Crystal Darter.[14] The species was not new to science, as the Crystal Darter (*Crystallaria asprella*) had been described as a new species in 1878 by Jordan.[15] But it was a newly discovered distribution record. In 1980, the distribution range of this species included a few Gulf Coast catchment areas and the Mississippi River drainage. When the Crystal Darter was discovered in the Elk River, the species was thought to be extirpated from the Ohio River drainage. The closest records (geographically and temporally) to the 1980 Elk River record included an 1888 record from the Muskingum River, near Beverly, Ohio, and an 1899 record from the Ohio River near Ironton, Ohio.

For years, I spent many nights in the Elk River hoping to catch a Crystal Darter. Pulling a bag seine in shallow to waist-deep water was a two-person job, so I was on one end of the seine and one of my graduate students was on the other end. It was exhausting at times, often with very little reward. But on the night of August 29, 2003, we pulled the bag seine, stretched it out onto an island shore, and shined the spotlight to see our catch. Two fish in the seine sparkled with reflected light like diamonds. A Rafinesque-like dance party commenced soon after, as we celebrated the catch of our rare crystals. Subsequently, we were unsuccessful at catching this fish until 2005, when two more specimens were seined. From 1980 to 2005, only 12 individuals were collected from the Elk River by us and others. After close examination of these Elk River specimens, and other museum specimens from the Ohio River drainage, I realized that the Ohio River population (with exception of the Wabash River) was different from Crystal Darter specimens from Gulf Coast drainages and other areas of the Mississippi River drainage. In 2008, I described the Ohio River population, previously recognized as the Crystal Darter, as a new species—the Diamond Darter, *Crystallaria cincotta*.

One year later, Dr. Jamie Geiger of the US Fish and Wildlife Service signed the "spotlight species" action plan for the Diamond Darter. The action plan detailed research needed to discover and document new information about this rare species, yet the primary problem was persistent—who in the world could consistently find this fish! On the night of September 13,

2011, we made a new discovery. During a break from bag seining, graduate student Crystal Ruble wandered off upstream with a spotlight. "I found one!" came her voice triumphantly rolling downstream out of the river fog toward myself and colleague Patrick Rakes. She had spotlighted a Diamond Darter, and just like that, a new sampling method was discovered. During the nights of September 13 and 14, we located 25 Diamond Darters, exceeding the total number of Diamond Darter observations during the previous 30 years (1980–2010)—actually, during the past 300 years. The "spotlight species" could be found using spotlights—who could have imagined it!

The spotlight sampling approach allowed for new discoveries on the type of river habitat used by the Diamond Darter. With the help of many graduate students, I discovered that Diamond Darters were frequently found in areas of the river just upstream of riffle habitat.[16] These areas, generally referred to as glide habitats, have relatively shallow depths, moderate water velocities, and a mix of sand, gravel, and cobble substrates (figure 20.2). Within glides, Diamond Darters were found resting on the river bottom at locations with average water column velocities of about 0.9 feet per second (27.4 cm per second) and bottom velocities of 0.5 feet per second (15.2 cm per second). The pieces of gravel, small stones less than 2.5 inches (6.4 cm) in size, and cobble, ranging from 2.5 to 10 inches (6.4–25.4 cm), reduce water velocities near the river bottom. The benthic locations of Diamond Darters had a high percentage of sand (52%), with lower percentages of gravel particles (26%) and cobblestones (20%).

Sedimentation of bottom substrates by clay or silt particles is common in many central Appalachian rivers. In our research, we noticed that stream sedimentation was minimal in the glide habitats used by Diamond Darters. Sand, as you are well aware, is composed of very small particles, often defined within a size range of 0.002–0.040 inches (0.05–1.00 mm). Even smaller particles, known as clay and silt, are less than 0.002 inches (0.05 mm). Excessive amounts of clay or silt are often detrimental to fishes and other aquatic life in central Appalachian streams. A common cause is the cumulative effect of landscape disturbances within a watershed, where the small dislodged particles enter streams by runoff from rainfall events. Small sediments fill interstitial spaces on stream bottoms, eliminating essential foraging, refuge, and spawning habitat for fishes and many other aquatic animals.

Pool habitats upstream of glides are scoured during high river flows. Sediments suspended in the water column begin to settle onto the stream

FIGURE 20.2. Nighttime photographs of Diamond Darters resting on the sand, gravel, and cobble bottom of a glide habitat in Elk River, West Virginia.

bottom during receding flows, but are swept from glide habitats owing to moderate water velocities in these areas. Diamond Darters need the clean-swept, sediment-free sandy glide habitats, as we discovered they often bury themselves in the sand. A blanket of sediment covering the river bottom would inhibit sand-burying behavior. Often Diamond Darters are completely buried in the sand, whereas sometimes they are partially buried with head exposed. Sand-burying behavior likely serves for predator avoidance and daytime refuge and as a method to prevent being swept downstream during high river flows. My observations have not supported sand-burying of the Diamond Darter as a lie-in-wait foraging tactic.

Subsequently, new discoveries were made on the reproductive biology and early life history of the Diamond Darter. These studies were overseen by Patrick Rakes and John R. Shute at their captive propagation facility (Conservation Fisheries) in Knoxville, Tennessee. The work was part of a graduate student's thesis.[17] In aquaria, spawning occurred in moderate-velocity areas of clean sandy substrate. Spawning behavior was observed, video recorded,[18] and summarized by Ruble et al.:

> Receptive females would swim into the territory and accept the male's vibration solicitation for spawning. During courtship males repeatedly swam directly in front of the female. Sometimes the male would position perpendicularly in front of the female and wag his tail, while at other times just quiver, then circle around and swim above her, almost resting on her back. The male would also approach beside the female and vibrate. Females would either swim away, or vibrate with the male, descending into the substrate. During this process the posterior 1/3–1/2 of the female's body descended into the sand and the male would be positioned along her side, tilted almost at a 45° angle with tail down and head in an upward position. Vibrations and burial into the sand usually coincided with a wide gape of the female's mouth.[19]

Success in spawning Diamond Darters in aquaria provided an opportunity to discover new information about eggs and larvae. As expected, the eggs were small, at 0.075 inches (1.9 mm) in diameter. Eggs hatched in about seven to nine days, and larvae were about 0.27 inches (6.9 mm) in length. In five days, larvae grew to about 0.38 inches (9.7 mm) in length. In ten days, larvae had further developed, and a most unusual new discovery was made—the presence of large, prominent teeth. This observation raised all sorts of questions, in particular, are Diamond Darter larvae predators of other fish larvae? The answer to this question is yet unknown. It also leads to another mystery: teeth are not prominent in adult Diamond Darters, so these teeth are either lost or become embedded as the larvae grow toward adulthood.

In 2013, the Diamond Darter was listed as an endangered species by the US Fish and Wildlife Service.[20] The only known extant population of this species is found in the lower 31 miles (50 km) of the Elk River, West Virginia. It would be interesting to know the distribution and rarity of this species during the days of Rafinesque, but spotlights were not available then, and collection of uncommon darters in large rivers was likely as difficult then as today. Currently, the cycle of new discoveries and new questions will continue, especially for the Diamond Darter, a singularly unique and rare fish species of the central Appalachians.

Fish Physics

On June 10, 1902, Osborne Reynolds presented the prestigious Rede Lecture at the University of Cambridge—it was titled *On an Inversion of Ideas as to the Structure of the Universe.*[1] The lecture was followed by his 1903 publication *The Sub-mechanics of the Universe.*[2] Reynolds, a British engineer and physicist, was an independent "big thinker" and a broadly productive researcher. His interests ranged widely through electricity, energy, fluids, friction, gases, gravitation, lubrication, motion, sound, turbulence, and viscosity. Reynold's research has contributed to our understanding of many subjects, including articular (joint) physiology, the cardiovascular system, bed formation in estuaries and rivers, gunsmithing, design and development of machinery, kinetic theory, lifeboats, meteorology, steam engines, and my personal favorite—the slipperiness of ice.

A particular area of interest for Reynolds was the properties of granular materials, which contributed to his ideas on the structure of the universe. Sand is an example of a granular material, where each particle is a solid but many particles together move like a fluid. Reynolds's studies of granular materials often involved experiments with sand, leading to a property he termed "dilatancy." In terms of granular soils, this property is important to understanding slope stability, building foundations, road and tunnel construction, as well as construction of burrows by animals. In 1885, Reynolds wrote,

> Without attempting anything like a complete dynamical theory, which will require a large development of mathematics, I would point out the existence of a singular fundamental property of such granular media which is not possessed by fluids or solids. On perceiving something which resembles nothing within the limits of one's knowledge,

a name is a matter of great difficulty. I have called this unique property of granular masses "dilatancy," because the property consists in a definite change of bulk, consequent on a definite change of shape or distortional strain, any disturbance whatever causing a change of volume and generally dilation.[3]

In other words, Reynolds demonstrated that external (shear) pressure applied to a mass of granular material, such as sand, causes the individual grains to rearrange so that the mass increases in volume. For sand, the increased volume is caused by an expansion of interstitial spaces between adjacent particles. Like Reynolds, I also have a fascination with granular material, particularly sand grains, but not from an engineering physics perspective or toward a resolved theory for the universe. My interest in sand is with its importance as habitat for psammophilous (sand-dwelling) fishes. Sand on the bottom of a riverbed can be loose, but can also be tightly compacted, where individual grains are held together by gravity. But as Reynolds pointed out, "If we walk on sand under water, it is always more or less soft, for the interstices can enlarge, drawing in water from above."[4]

A particularly fascinating psammophilous fish of the central Appalachians is the Eastern Sand Darter (figure 21.1). Individuals of this species are small, less than three inches long (7.6 cm), with streamlined elongate bodies, low profiles, somewhat pointed but terminally rounded snouts, and relatively large pelvic and pectoral fins. The Eastern Sand Darter (hereafter shortened to Sand Darter) is a benthic fish without a swim bladder, living on the river bottom.[5] A swim bladder, a gas-filled sac that provides buoyancy, is present in most pelagic fishes that live primarily in the water column (above the river bottom).

Sandy areas of river bottoms are the preferred habitat for the benthic Sand Darter, where the sizes of these habitats can range from small patches to large expanses.[6] These sandy areas are often relatively flat, sometimes dome-shaped like a dune, and located in depositional areas or lower reaches

FIGURE 21.1. An Eastern Sand Darter, 2.4 inches (6.1 cm) in length.

of rivers, where sand grains from the watershed are transported downstream and deposited onto the riverbed.[7]

Water-current velocities, or the speed of water moving downstream, can be slow at sand-depositional areas during low river flows, but can increase substantially in strength at higher flows. Within the sandy areas, boulders, cobbles, or large woody debris are often not available for Sand Darters to hide behind or underneath. Seemingly, a lack of habitat structure would cause concern for individual survival, as structural hiding places are not available for avoiding predators. For benthic river fishes, areas immediately downstream of habitat structure provide refuge from water-current velocity, preventing fishes from being flushed downstream during higher river flows. With a lack of habitat structure on these sandy areas, Sand Darters are exposed to moderate to high water velocities during periods of higher river flows. This brings forward an interesting question—how do Sand Darters maintain their benthic position without being flushed downstream during periods of higher river flows?

The ability of a riverine fish to maintain a fixed position is called "station-holding."[8] I do not know of any research studies on station-holding of the Sand Darter, but studies have focused on other central Appalachian fishes, including species of minnows, suckers, trout, sculpins, and darters.[9] As an analogy, I have experienced the difficulty of station-holding when snorkeling in fast moving rivers, which I describe below in real time:

> Completely submerged with my head upstream, I feel the forces of the water currents on my body. Vortices of turbulent water, shed from the edges of my facemask, gurgle loudly next to my ears. The pressure of the water current on my head and shoulders and the friction of the water along my body effectively drag me slowly downstream. I am taxed to swim against the current, so I opt for some creative counter-measures. I grasp some loose cobblestones, but this only slows my downstream descent. To successfully hold station, I reposition myself in an eddy on the downstream side of a large boulder. Next, I move laterally in the river to a swift area with a sandy bottom, but I am not able to hold station as I have nothing to grip or to hide behind. I am enviously fascinated by the ability of the benthic fishes surrounding me, particularly the Sand Darters, to successfully hold station, while I struggle to do so.

Station-holding of benthic fishes is aided by an interesting phenomenon—water-current velocity near the river bed is often much lower than in the water column. Consequently, a benthic fish would be exposed to lower velocities than those experienced by a pelagic one. One of my graduate students noted this in her study of habitat use of Sand Darters. On average, water-current velocity near midcolumn was 0.33 feet per second (10.1 cm per second), but that at about one inch (2.5 cm) above the bottom was reduced in half at 0.16 feet per second (4.9 cm per second).[10] Slower water-current velocities near the river bottom represent the "boundary layer effect."

Within the benthic boundary layer, water-current velocities of laminar or turbulent flow decrease near the river bottom (figure 21.2). A thin strip of the boundary layer, which adheres to a smooth river bottom, is known as a viscous sublayer. When flow is laminar, water slides over the surface in smooth layers with linear streamlines. Depending on the texture of the river bottom, flows in the boundary layer above the viscous sublayer can range from smooth turbulent to rough turbulent.[11] In context, turbulence is chaotic or erratic water motion. A sandy river bottom has a relatively smooth surface, so the boundary layer flow likely includes a viscous sublayer underneath a smooth turbulent layer. In contrast, a gravel or cobble bottom would produce a boundary layer with more turbulent (or chaotic) flow directions, and the viscous sublayer would be disrupted. Benthic fishes, such as the Sand Darter, benefit from the boundary layer effect, as the lower water-current velocities and lack of rough turbulence near the sandy river bottom aid in station-holding during low to moderate river flows.

Flows ranging from laminar to turbulent can be calculated, which takes us back to Osborne Reynolds, who studied fluid flows in a pipe. In a landmark 1883 paper,[12] Reynolds referred to turbulent flow as "sinuous," "irregular," or "unsteady."[13] In general, Reynolds calculated the ratio of inertial force to viscous force; this dimensionless number is called the "Reynolds number."

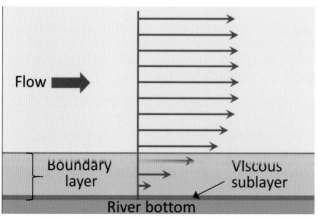

FIGURE 21.2. Water velocity profile above a smooth river bottom emphasizing the benthic boundary layer and its viscous sublayer. Arrow lengths are proportional to water velocities.

Basically, the inertial force is the force associated with the momentum of the fluid, where momentum is influenced by the density and velocity of the fluid. The viscous force is a force owing to friction between fluid layers, which results in a resistance to fluid flow. Important to our interest in station-holding of the Sand Darter, low Reynolds numbers describe laminar flows, whereas higher Reynolds numbers describe turbulent flows. In rivers with high flows, high Reynolds numbers and turbulent flows are the norm, even over the relatively smooth sandy habitat of the Sand Darter. High-velocity turbulent flows make it extremely difficult for a Sand Darter to hold its station.

Similar to the boundary layer of the river bottom, a stationary object submerged in flowing water, such as a streamlined Sand Darter, experiences a boundary layer along its body. The boundary layer phenomenon was documented by Ludwig Prandtl,[14] a German engineer, a physicist, and the father of modern aerodynamics.[15] Prandtl emphasized that a moving fluid adheres to the body of a moving or stationary object, but can also separate from that object. In 1904, Prandtl wrote: "By far the most important question of the problem is the behavior of the fluid on the surface of a solid body. A satisfactory account can be determined for the physical phenomena in the boundary layer between the fluid and the solid body if one assumes that the fluid adheres to the surface, where the velocity equals zero or equals the velocity of the body."[16] Prandtl illustrated the viscous sublayer wrapping around a cylindrical object within a flow field. Streamlines of the boundary layer wrap around the sides. The laminar sublayer can separate from the surface, where a wake of turbulent vortices occurs on the downstream side (figure 21.3).[17]

For a Sand Darter, the ability to hold station on a sand flat in a swift river requires either swimming against the flow or some sort of countermeasure. The Sand Darter is not a particularly strong swimmer. Station-holding is

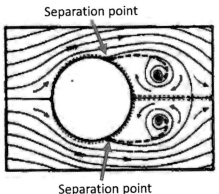

Separation point

FIGURE 21.3. Streamlines around a cylinder (adapted from Prandtl 1904), showing the point where the viscous sublayer separates from the object, creating turbulent vortices, called eddies, on the downstream side.

Separation point

aided by its streamlined low-profile body, which has little resistance to moderate velocity flows when oriented upstream—a common behavior of benthic fishes, known as rheotaxis.[18] The Sand Darter, when holding station in water current, has a boundary layer over its body surface. A viscous sublayer attempts to adhere to the darter's body, but may be disrupted by the rough surface of the scales (more on that later). For the streamlined Sand Darter to experience low drag, the boundary layer around its body must remain adhered to the entire length of the body surface, resulting in a narrow downstream wake (figure 21.4).

From a physics perspective, resistance is conceptualized as drag. Drag is commonly separated into pressure and frictional drags. Profile drag, the sum of pressure and frictional drag, can dislodge a Sand Darter from its station-holding position. Despite a streamlined body, the fish's rounded head produces a pressure drag. It would be reasonable to assume that a fish with a sharp, pointed snout would have less drag, but the Sand Darter's rounded snout allows it to turn its head slightly without creating a lot of extra drag. Contrastingly, the sublayer separates from the turned head of a sharp-snouted fish, causing drag and difficulty with station-holding. This separation of the sublayer and disruption of the streamlines can be seen when a fish begins to turn sideways in the current, where pressure drag is created by two separations of the sublayer, called shear layers, leaving a wake of vortices downstream (figure 21.4).[19]

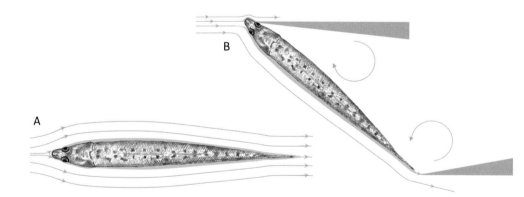

FIGURE 21.4. Streamlines of flow around the upper body of an upstream-facing Sand Darter (A). The viscous sublayer adjacent to the body is shown in gray. When an individual's body is positioned diagonal to current direction (B), separation points of the viscous sublayer occur near the head and tail, producing a downstream wake and hindering the fish's ability to hold station. This example assumes adpressed pectoral fins. Concepts based on Anderson (2005).

Friction drag, also called skin friction drag, occurs as the layers of water pass along the side of the fish's body. Consider a sliding deck of cards as moving layers of fluid.[20] For a deck of cards sliding across a tabletop, friction occurs between the bottom card and the tabletop, as well as between each card in the deck. The velocity of the cards increases from the bottom to the top card. Although many fishes have a noticeable coat of mucus on their bodies, which may reduce drag,[21] a portion of the Sand Darter's body surface is slightly rough to the touch, owing to the presence of small spines (ctenii) on the posterior margin of ctenoid scales (figure 21.5).[22]

I wonder about the friction drag created by this scaled surface roughness. Intuitively, it would be reasonable to assume that the rough surface would cause separation of the boundary layer, more turbulence, and increased drag. Consider a non-fish example of a dimpled golf ball, where drag is reduced by as much as 50% as compared to a ball with a smooth surface.[23]

FIGURE 21.5. An illustration of a ctenoid scale of an Eastern Sand Darter (0.9 mm width) with small spines (ctenii). The posterior portion of the scale with the pointed ctenii is exposed, contributing to a partial body surface with a rough texture. The remaining section of the scale with the concentric rings (annuli) is embedded.

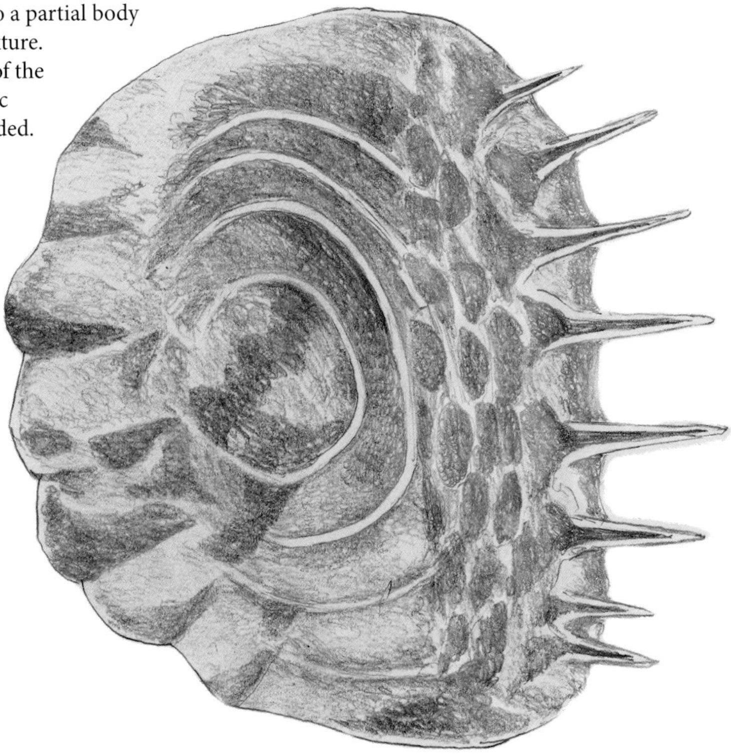

For the golf ball, the boundary layer separates at each dimple, causing a bubble of turbulence, but then reattaches to the surface at a higher momentum, thus delaying permanent separation of the boundary layer and reducing drag. Possibly, a similar golf-ball effect on the boundary layer occurs over the rough scaled surface of the Sand Darter's body, reducing friction drag and aiding benthic station-holding. Evidence for this rough-surface effect has been documented for other fish species. Consider the rough skin of many species of fast-swimming sharks, where denticles (shark scales) are known to reduce drag.[24] In another fish example, similar to our Sand Darter, a rough surface caused by ctenoid scales has been hypothesized to reduce drag on bodies of Bluegill Sunfish.[25]

A Sand Darter may be able to increase its station-holding ability by simply changing its body posture. When facing no or low velocities, the Sand Darter often props up the front end of its body with the pelvic fins or the tips of its pectoral fins, a posture known as "fin standing." In higher velocities, an inclined body in this propped position would not allow station-holding. To improve station-holding in higher flows, a Sand Darter may change body posture similar to that found for the Fantail Darter, where downstream slippage was reduced by a lowered and downward-angled head position combined with an arched back and contracted dorsal and caudal fins.[26] Also, this posture could be combined with friction of pelvic or pectoral fins in contact with the sandy substrate.[27]

The pectoral fins, located on the sides of the body, may also aid in station-holding. Consider a comparison between the pectoral fin and an aerodynamic airplane wing. An airplane wing or airfoil produces lift, based on a principle from the work of Daniel Bernoulli, a Swiss mathematician and physicist. The Bernoulli principle states that pressure is reduced with increased velocity. For an asymmetrical airplane wing, with pronounced curvature over the top, the velocity of flow over the top is faster than that on the bottom, resulting in positive lift because of less pressure on the top side (figure 21.6). If a benthic fish positioned its pectoral fin to mimic an airplane wing, then it would be lifted off the bottom of the river, losing its ability to maintain station. However, if the pectoral fin is tilted downward, where the lower portion is projected forward and the upper part is cupped upwards, then a negative lift is possible.[28] Also, the cupped fin may cause separation of the boundary layer, where a clockwise vortex behind the fin moves water in an upstream direction, assisting station-holding.[29]

Although the pectoral fin likely plays a hydrofoil role in station-holding, its effect is likely minimal compared to that of fin friction.[30] The pelvic and

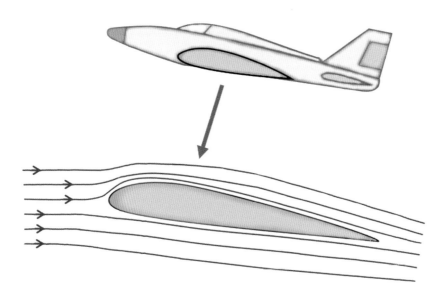

FIGURE 21.6. Positive lift of an airplane wing, a type of airfoil, is explained by the Bernoulli principle, where increased velocity (crowded streamlines) on top of the asymmetric airfoil causes less pressure. Presumably, a fish could produce a similar hydrofoil effect with its pectoral fin. A negative lift could result from slightly cupping the fin and slanting the fin front downward.

pectoral fins of the Sand Darter are often in contact with the sandy bottom of the river, deterring downstream slippage. In an arched-back posture the caudal and anal fins may also be in contact with the substrate, providing further friction against displacement.

Despite all the interesting possibilities that physics can play in station-holding, a Sand Darter may simply flee higher flows. At some critical point of increasing current velocity, a benthic fish will start to slip downstream. When experiencing this slip speed, a Sand Darter may move toward the river margin, where velocities are lower. The fish also has another option, which is perhaps one of its most interesting behaviors—sand-burying. In 1877, David Starr Jordan and Herbert Edson Copeland documented sand-burying behavior of an aquarium-held Eastern Sand Darter:

> There for days we watched it closely, only to learn that it could bury itself with great celerity, for it was never caught in the act. Our patience was at last rewarded, however, for as we came out to breakfast one morning it put its nose, that we now know has a tip nearly as hard as horn, against the bottom, stood nearly straight on its head,

and with a swift beating of its tail to the right and left was in less than five seconds completely buried. The sand had been violently stirred, of course, and just as it had nearly settled, probably in less than half a minute, its nose was pushed quietly out and, settling back, left the twinkling eyes and narrow forehead alone visible.[31]

Presumably, sand-burying could serve a role similar to station-holding,[32] as the fish would maintain position by burying itself in the sand. I have observed Sand Darters in aquaria for many hours. The burying process is usually complete in less than a second.[33] I have observed two forms of sand burial. First, as noted by Jordan and Copeland, sometimes an individual is only partially buried with head exposed (figure 21.7).[34] Second, the bodies of some individuals are completely covered, often remaining there for hours. Sand-burying is a fascinating behavior of our little Sand Darter and, in addition to a role similar to station-holding, may have other explanations, such as predator avoidance,[35] lie-in-wait foraging,[36] or resting behavior.[37]

In summary, physics predominates explanations for station-holding of the Sand Darter, including the streamlined body, the benthic and body

FIGURE 21.7. Can you find the Sand Darter in this photograph? The darter's body is buried in the sand with an eye exposed.

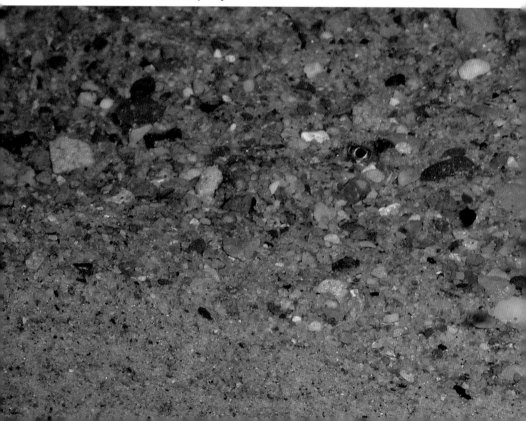

boundary layers, friction of fins on the river bottom, and the possibility of negative lift by pectoral fin hydrofoils. Perhaps sand-burying is one of the more interesting explanations, as it likely serves a role in station-holding. During my homework for this story, I discovered that many physicists have furthered the field of fluid dynamics, with findings relevant to station-holding in benthic fishes, including Archimedes, Jean le Rond d'Alembert, Leonhard Euler, Lord Kelvin, Lord Rayleigh, Horace Lamb, Claude-Louis Navier, Sir Isaac Newton, George Stokes, and others.[38] Also, many of Prandtl's students made substantial contributions to the field; Heinrich Blasius, Theodore von Kármán, and Herman Schlichting. I wonder, would they have shared my interest in the Sand Darter? I suspect that Osborne Reynolds, if alive today, would be quite fascinated with the physics behind the Eastern Sand Darter's station-holding and sand-burying acts. Reynolds's concept of "dilatancy" likely plays an important role in sand-burying.[39] Admittedly, station-holding or dilatancy explanations have not been tested for the Sand Darter, but my hope is that this story will enlighten the inner physicist of an aspiring ichthyologist—encouraging further studies on this singularly fascinating psammophilous fish of central Appalachian rivers.

Tippecanoe Is Tiny Too

In the 1800s, naturalists were often physicians or politicians, an unusual trilogy by today's standards. Consider DeWitt Clinton, a naturalist and politician who was a United States senator, mayor of New York City, and the sixth governor of New York. As a naturalist, Clinton described the Spottail Shiner (*Notropis hudsonius*) in 1824,[1] a species of minnow native to the Atlantic slope of the central Appalachians. Clinton also provided specimens to the naturalist Constantine Samuel Rafinesque, which resulted in the 1818 description of a new genus, *Notropis,* and a new species, *Notropis atherinoides* (Emerald Shiner).[2] Rafinesque described and named the plant genus *Clintonia* in Clinton's honor. The Blue-Bead Lily (*Clintonia borealis*) and Clinton's Lily (*Clintonia umbellulata*) are both late-spring bloomers of central Appalachian forests. In 1832, Rafinesque wrote, "Of all the New Genera of Plants which I claim to have established and well named, to few am I more partial than to the beautiful G. CLINTONIA which I published in 1817 in America and in 1819 in France (50 N. G. Journal phys.) of the natural tribe of Asparagides; which I dedicated to my worthy friend Dewitt Clinton, an eminent Philosopher, Naturalist and Statesman."[3]

DeWitt Clinton ran for president in 1812 but was defeated by James Madison. Clinton was extremely tall for the time at six feet, three inches (191 cm), where James Madison was the shortest ever presidential candidate at five feet, four inches (163 cm). I find this interesting because many people assume that taller presidential candidates usually win elections over shorter opposition, similar to the common belief that most larger-bodied animal species have a competitive advantage over smaller-bodied species.

As further example, consider the tremendously popular 1840 campaign song "Tippecanoe and Tyler Too." Whig Party candidate William Henry

Harrison (nicknamed Old Tippecanoe) was running for president with vice presidential candidate John Tyler. The five-foot, eight-inch (173-cm) Harrison, who was relatively tall for the time, was nicknamed for his heroism at the 1811 Battle of Tippecanoe near the Tippecanoe River in Indiana. The song's chorus campaigned for support of the heroic Harrison over his smaller-sized opponent, the five-foot, six-inch (168 cm) Martin Van Buren, referred to as "little Van":

> Tippecanoe and Tyler Too
> Tippecanoe and Tyler Too
> And with them we'll beat little Van, Van, Van
> Van is a used up man
> And with them we'll beat little Van

The slightly taller Harrison successfully defeated Van Buren in the 1840 race. Although there are many examples of taller presidential candidates winning elections, there are almost as many examples of elected shorter candidates, such as in the 1836 election, when Van Buren defeated Harrison. Size does not seem to matter.

In nature, individuals are obviously not running for political office, but competition is a common occurrence. Individuals compete for many resources, including space for refuge, foraging and spawning areas, and directly for food. Like many, you might assume that larger individuals have a competitive advantage over their smaller counterparts. Can a fish species with small-bodied individuals find success when surrounded by species of larger-sized individuals? Consider the tiny individuals of a central Appalachian species, the Tippecanoe Darter, which rarely exceed a length of 1.5 inches (38 mm) (figure 22.1). Darters are a diverse group of small-sized fishes of the family Percidae,[4] with the largest darter species reaching a maximum length of nearly 8 inches (20.3 cm).[5] Members of a fish species are often considered as miniature when their maximum length does not exceed about 1.25 inches (32 mm);[6] the maximum length of the Tippecanoe Darter is close to miniature status.[7] Like William Henry Harrison's nickname, the Tippecanoe Darter's name comes from the Tippecanoe River in Indiana.[8] This darter has a much wider distribution than just Indiana, including medium- to large-sized rivers of the central Appalachians within Ohio, Pennsylvania, and West Virginia. Very likely, the inquisitive nature of your naturalist mind is already asking questions about this small fish; in particular, how can the small-sized Tippecanoe Darter be successful in a large river?

Assumptions made about the popular phrase "survival of the fittest" partly promote a misconception, as many people reason that the "fittest" must be the

FIGURE 22.1. Magnified views of the male (*top*) and female (*bottom*) Tippecanoe Darter. The male and female are approximately 1.4 and 1.2 inches (36 and 30 mm) in length, respectively.

largest and strongest individuals. Not true. Evolution toward smaller body size has occurred in many animal groups.[9] Consider Spring Peepers and Upland Chorus Frogs, examples of small tree frogs of the central Appalachians. The Ruby-throated Hummingbird is an excellent example of a small bird that is very popular among bird enthusiasts of the central Appalachians. One of my favorite central Appalachian miniatures is the Winter Wren, a small brown bird with an upturned tail that I see occasionally along mountain streams. Small-sized fishes are also common, such as the madtom catfishes (see chapter 11, "Madtom Miniatures"). Within the family Percidae, darters are small-sized fishes, but the miniature Tippecanoe Darter takes it down to a lower level of littleness. One might surmise that small fish would live in small streams, but despite a small body size, Tippecanoe Darters are often found in large rivers. From a fish-use perspective, large rivers are sectioned into smaller habitat units. Three general habitat types in rivers include pools, runs, and riffles. Pools represent deeper habitat with slow water velocities, often with small substrates, such as silt and sand, covering the riverbed. Runs are moderate-depth and moderate-velocity habitats with a mixture of small and larger substrates. The Tippecanoe Darter typically inhabits riffles, which are the shallower, higher-gradient sections with relatively fast water velocities and a wide range of substrates, such as small and large gravel, cobbles, and boulders.

The proportions of gravel, cobble, and boulder substrate vary between riffles, and the distributions of these rock sizes vary within riffles. The Tippecanoe Darter often associates with gravel substrates of riffles.[10] Interestingly, all sizes of gravel are not the same, which is something I learned at a young age. One of my prized possessions was a gravel shooter (slingshot), carefully crafted by my grandfather from a pronged limb of a maple tree, a couple of rubber strips, and a leather pouch from the tongue of an old boot. Gravel-sized rocks were abundant along the railroad tracks, as were the tall, long-stemmed plants commonly called Joe-Pye weed. Just the right-sized piece of gravel was accurate and deadly at toppling the Joe-Pye weeds. It seemed reasonable to me to knock these plants over before the annual herbicide applications along the railroad right-of-way. A small-sized grain of gravel did not have the amount of mass needed to get the job done. A large piece of gravel, if it didn't bust my knuckles when passing through the prongs, produced a lob with low velocity and less accuracy.

A range of medium-sized pieces of gravel appears to be the preference for the Tippecanoe Darter. Similar to my gravel-shooter story, there is a Goldilocks effect with the Tippecanoe Darter's rock-size selection. The effect owes to the size of interstitial spaces; that is, crevices between the rocks. The Tippecanoe Darter's smaller size allows it to fit into small crevices between pieces of gravel, which likely provide refuge from water velocities of riffle habitat and a safe haven from predators. For small grains of gravel, crevice sizes are too small for the Tippecanoe Darter. Crevices between large pieces of gravels provide space for the darter, but are not the best fit for shelter and refuge. The medium-sized pieces of gravel are just right.

At first thought, one might assume the faster water velocities of riffles would wash the small-sized Tippecanoe Darters downstream. Velocities near the water's surface in a riffle are much faster than those near the stream bottom, a phenomenon influenced in part by the boundary layer effect (see chapter 21, "Fish Physics"). Also, larger rocks upstream often create an eddy effect, giving the Tippecanoe Darter a velocity shelter in gravel-deposition areas. Additionally, by being benthic and having access to spaces between pieces of gravel, the Tippecanoe Darter is able to hold station and avoid downstream displacement.

A big fish in a small pond has its advantages, particularly if you want to get noticed. A small fish in big water has the opposite effect—you go un-noticed. Going unnoticed is particularly advantageous to the Tippecanoe Darter, as it provides protection from predators. Also, due to large size, many predators are excluded from fast-velocity, shallow riffles of large rivers. Fur-thermore, by having access to interstitial spaces between pieces of gravel, the Tippecanoe Darter is able to avoid most predators.

The crevices between gravel pieces, which give the Tippecanoe Darter a refuge from velocity and predators, also provide a hideout for tiny aquatic insect larvae, often called macroinvertebrates, the favorite food of the Tippe-canoe Darter. Small fishes eat small foods. One diet study reported larval ratios of 57% Trichoptera (caddisflies), 39% chironomids (midges), and 4%

Ephemeroptera (mayflies).[11] The Tippecanoe Darter likely competes with other darters for food, but with exclusive access to narrow crevices between medium-sized pieces of gravel, this small fish has a foraging advantage in its preferred habitat.

The Tippecanoe Darter is a short-lived species. Initially, it was thought that it was an annual species, but further observations have found that individuals live two years. A "live fast, die young" lifestyle requires sexual maturity at a young age, as individuals spawn at an age of one.[12] Many darters spawn during spring, but the Tippecanoe Darter spawns in late June or July. Late spawning is likely beneficial, as river flow fluctuations are less in summer than that of springtime, and large flow fluctuations often result in failed spawning attempts. Given a short life span, the Tippecanoe Darter cannot afford spawning failure in two consecutive years.

The Tippecanoe Darter has a track record of being a successful small fish in a large river. But population extirpations have occurred across its geographic range. These lost populations are reflected by its fragmented distribution range (figure 22.2).[13] Fragmented ranges often reflect fragmented habitat, where sections of continuous habitat have been degraded and are no longer hospitable to fishes. Stream sedimentation is likely an important habitat degrader in the Tippecanoe Darter's story. Stream sedimentation often follows land disturbance, where clay or silt-sized particles of less than 0.002 inches (0.05 mm) enter streams by runoff from rainfall events. These small particles are deposited onto stream bottoms, filling interstitial spaces and eliminating essential foraging, refuge, and spawning habitat.

The next time you are driving along or across a large river, give some thought to the diversity of fishes. Nature's diversity is saturated with small-statured individuals. If you find yourself reasoning that large rivers are only for large fishes, then consider the Tippecanoe Darter—a small-sized testament to the trials and tribulations of being tiny in a large river world.

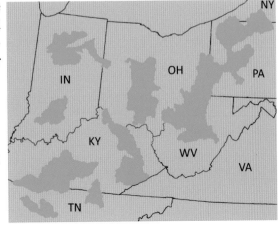

FIGURE 22.2. A watershed-based map of the fragmented distribution range of the Tippecanoe Darter. The map is based on data compiled for a 2018 species status assessment by the US Fish and Wildlife Service.

Darter of Darters

In the foreword of his 1922 autobiography, the naturalist and ichthyologist David Starr Jordan opens with a modest understatement "For half a century the writer of these pages has been a very busy man."[1] Jordan's contributions to science were beyond stellar, as he excelled as teacher, mentor, researcher, and prolific publisher. Most current ichthyologists can trace their academic lineages back to Jordan or his close associates. Carl Leavitt Hubbs (Jordan's last student) noted that Jordan produced 2,017 publications, including 645 contributions to ichthyology.[2] Jordan's sheer number of ichthyology papers is hard to comprehend, yet he also found time to author popular scientific essays. An anonymous reviewer in the *Popular Science Monthly* (1888) referenced Jordan's essays as "of a popular character, and written in a pleasing style, though without sacrificing scientific accuracy."[3] As an example, I quote some descriptive words on the Fantail Darter, extracted from one of his essays on darters, published in 1887: "The darter of darters is the Fan-tail (*Etheostoma flabellare* Rafinesque). Hardiest, wiriest, wariest of them all, it is the one which is most expert in catching other creatures, and the one which most surely evades your clutch. You can catch a weasel asleep when you can put your finger on one of these. It is a slim, narrow, black, pirate-rigged little fish, with a long pointed head, and a projecting, prow-like lower jaw."[4]

The Fantail, common within the central Appalachians, lives in small creeks and large rivers. Jordan emphasized the shape of the tail, from which the darter gets its name: "Its fins, especially the broad fan-shaped caudal, are likewise much checkered with spots of black."[5] Maturity is reached by age two, and most individuals do not live longer than five years.[6] Adults rarely exceed five inches (7.6 cm) in length. Adults have a series of vertical bars of pigment on the lateral sides of the body. A dark blotch of pigment sits behind

FIGURE 23.1. Breeding male Fantail Darter, about 2.2 inches (5.6 cm) in length, from the Ohio River drainage. Note the egg mimics on the distal edge of the first dorsal fin.

FIGURE 23.2. Gravid female Fantail Darter, about 2.0 inches (5.1 cm) in length, from the Ohio River drainage.

the head, just above the base of the pectoral fin. Blotches or bands of dark pigment are present in front and behind the eye, with some individuals having a dark spot below the eye (figures 23.1, 23.2).

The Fantail is sexually dimorphic, meaning that males and females differ in appearance. Males are larger than females, particularly in the size of the head and the pectoral, pelvic, and anal fins.[7] Breeding males become "slimy" from an increase in mucus production.[8] In 1887, Jordan emphasized the "black" coloration of Fantails, most likely describing the black head of a breeding male, a color character that fades fast after its captured, particularly when you are trying to pose a fish for a photograph (figure 23.1). For many species of darters, males differ from females in developing bright breeding colors, but this is not the case for the Fantails, as emphasized by Jordan: "It carries no flag, but is colored like the rocks, among which it lives."[9]

Perhaps the most unique difference between the Fantail's sexes is the presence of fleshy knobs on the tips of the dorsal fin spines in breeding males. Jordan emphasized this unusual attribute of the dorsal fin: "The spines of the dorsal fin are very low; and each of these in the male ends in a little fleshy pad of a rusty-red color, the fish's only attempt at ornamentation."[10] Geographic variation in coloration of the fleshy knobs range from Jordan's rusty red to orange, golden yellow, or amber. More recently, researchers have proposed that these fleshy knobs look a lot like fish eggs (see figure 23.1).[11] This dorsal-fin egg-mimic hypothesis is quite perplexing at first, particularly when you ask the question why any respectable male fish would have "egg mimics" on its dorsal fin. The answer to this question lies in understanding the spawning behavior of the species.

The Fantail Darter spawns during spring to early summer within shallow and moderate water-velocity sections of streams. A favorite spot for spawning is the underside of a rock. The male selects the spawning rock, and the female selects the male. Female choice is a common strategy of mate selection in fishes, but often the choice is influenced by the brightly colored appearance of the male. But Fantail males are not brightly colored, leading to an initial speculation as to how a female chooses a mate.

To understand egg mimicry and mate choice we need to first consider the species' sexual actions. During the breeding season, often March–June, a male Fantail establishes his breeding territory in a cavity under a rock, often a flat rock. Spawning behavior was described in detail from the doctoral research of Howard Elliott Winn, published in 1958, an ichthyologist whose academic lineage traces back to David Starr Jordan via Robert Rush Miller and Carl Leavitt Hubbs. Winn wrote: "When a female enters a male's territory he typically comes out to meet her and swims in advance to the rock nest. The female moves under the rock and if she does not turn upside down she is nudged or chased away. The presence of the female upside down under the rock stimulates movements of the male which culminate in fertilization of the eggs. The male swims upside down beside her and prods her if she stops her movements over the rock. One egg is deposited at a time."[12]

The female is often indecisive, entering and leaving the nest cavity with the male several times before remaining to spawn.[13] Presumably, repeat visits to the nest give the female an opportunity to determine if nest conditions are adequate. Winn illustrated opposing head-to-tail body positions of the male and female during pre-spawning and spawning (figure 23.3).[14] During pre-spawn, the male's body, with fleshy-tipped dorsal fin spines erected toward the ceiling of the nest cavity, is positioned adjacent to the upside-down female's body. This position is necessary for attaching eggs to the underside of the rock. Each egg attached to the rock ceiling of the nest cavity by a female is simultaneously fertilized by the male. A female may spawn in the

FIGURE 23.3. Nest cavity spawning behavior of the Fantail Darter, illustrating opposing positions of the male (M) and upside-down female (F). The male is horizontal during prespawn with dorsal fin erect toward the cavity ceiling, and upside down during egg fertilization. Adapted from Winn (1958), courtesy of the Ecological Society of America. Used with permission of John Wiley and Sons, from Howard E. Winn, "Comparative Reproductive Behavior and Ecology of Fourteen Species of Darters," *Ecological Monographs* 28, no. 2 (1958): 155–91. Permission conveyed through Copyright Clearance Center, Inc.

nest cavities of multiple males, laying as many as 45 eggs during a spawning event. Also, several females contribute to the total number of eggs in a male's nest, which can result in a single-layer cluster of over several hundred eggs. Egg size ranges between two and three millimeters in diameter.

A male Fantail provides parental care of eggs in several ways. First, the male's presence and territorial behavior within the nest cavity reduces egg predation. The male remains with the eggs until the time of hatching, which, depending on water temperature, ranges from 14 to 35 days. Second, nesting males keep the eggs and nest cavity clean by fanning or brushing the eggs using the fleshy knobs of the first dorsal fin or the outer margin of the second dorsal fin.[15] Finally, the sliminess of the mucus-producing breeding male, particularly on the nape (the region of the back between head and first dorsal fin), may provide antimicrobial properties, as the male applies this mucus to the eggs to reduce fungal and bacterial infection.[16] The fleshy knobs on the first dorsal fin do not produce the mucus.[17] Experimental removal of the male immediately after egg deposition and fertilization coincides with an increase in fungal infection and mortality of the eggs.[18] What female would not be attracted to a nest-guarding male, one that keeps a clean house while protecting her eggs from predators, fungus, and bacteria? But in a cart-before-the-horse scenario, how can a female predict a male's nest-guarding, egg-fanning, and egg-protecting abilities prior to spawning?

A perplexing part of the paternal play is that a breeding male Fantail may eat some or all of the eggs in the nest,[19] a behavior known as filial cannibalism.[20] Males with small egg clusters will sometimes consume all the eggs in the nest, whereas large egg clusters generally experience partial cannibalism. Possibly, nest-guarding males maintain a healthy body condition by eating some of the eggs, given that food availability may be limited within the nest cavity. Small egg clusters may not be a worthwhile reproductive investment for males, possibly explaining instances of total cannibalism.[21] Additionally, females eat eggs from the nest. Egg cannibalism may improve female body condition, promoting future spawning success, or a female may improve the probability of survival of her own eggs by eating the eggs of others.

In an interesting twist to the Fantail Darter's sexual saga, females often independently choose to mate with males that already have eggs in the nest.[22] An alternative hypothesis is that a female's mate choice may not be independent, but rather the female may observe and "copy" another female's mate choice,[23] a strategy that also results in spawning with an egg-guarding male. The female choice of mating with an egg-guarding male may help answer questions relative to how a female can predict a male's nest-guarding and egg-fanning abilities, as well as the probability of filial egg cannibalism. First, a female may recognize an egg-guarding male as a successful and protective parent. Second, a female can observe or assume a male's egg-fanning and house-cleaning behavior relative to the condition of the existing eggs in the nest. Finally, a larger mass of eggs is often associated with an increase in parental care and egg survival.[24] A female may recognize that the presence of other eggs will reduce the probability of predation or filial cannibalism on her eggs.[25]

The female strategy of choosing an egg-guarding male leads us to an explanation for egg mimicry. When the male erects his dorsal fin while in the nest cavity, the fleshy knobs on the tips of the dorsal fin spines may appear as eggs attached to the cavity's ceiling (figure 23.4). Given that females use egg presence as a criterion for mate choice, then the female may be tricked into believing that eggs are present in the nest. In an experimental study with a small sample size, Roland A. Knapp and Robert Craig Sargent let females choose between males with egg mimics and males whose egg mimics had been surgically removed.[26] Of 17 spawning events, 13 occurred with males with egg mimics, suggesting that females prefer egg-mimic males.

FIGURE 23.4. A breeding male Fantail Darter showing the fleshy knobs of the dorsal spine tips as "egg mimics."

In my experiences with studies on fish behavior, truth is often elusive, and observational data often provide supporting evidence for two or more alternative hypotheses, as is the case with the Fantail's reproductive behavior. For example, males appear to guard the nest, but possibly they are just waiting around to spawn with subsequent females.[27] Experimental evidence supports a female preference to spawn in nests already containing eggs, but there are multiple explanations. The female may see existing eggs in a nest as evidence of a successful egg-guarding male. The female may recognize that the survival of her eggs increases in the presence of other eggs, but this seems inconsistent with female cannibalism of eggs in the nest. Filial cannibalism would seemingly reflect poorly on a male's parental skills, but alternatively it may actually maintain the male's body condition and improve his ability to protect the remaining eggs.

The fleshy knobs on the first dorsal fin appear to mimic eggs, which presumably increases male mating success. This is a reasonable hypothesis given that females prefer to spawn in egg-containing nests. Alternatively, the fleshy knobs may just serve as a protective covering of the spine tips, which reduces puncturing of eggs in the nest—although you must admit, the fleshy knobs do look a lot like eggs. I suspect that David Starr Jordan would have been quite fascinated with the egg-mimicry hypothesis, particularly with respect to the Fantail, the darter of darters.

Epilogue

My calling, as a scientist and an educator, has been to study and learn about fishes. Along the way, I have developed an appreciation of my scientific forefathers, those old-school naturalists from previous centuries who have contributed greatly to our current knowledge. I also carry a responsibility to share my research findings, which I have accomplished, in part, through mentoring and working closely with my graduate students. Our experiences have covered a wide range of topics, many of which have focused on central Appalachian fishes. Our research efforts include reading reams of scientific literature, collecting and analyzing data, and publishing research findings in scientific journals. Few people read scientific journals. Many of the old-school naturalists submitted papers to journals but also published stories and essays about nature, presumably to reach a wider audience. My inspiration for writing this book is based on the idea that many people are curious about nature and would enjoy learning more through reading.

The 23 chapters of this book represent a collection of stories on central Appalachian fishes, with an emphasis on ecological themes and contributions from old-school naturalists. As stated in the preface, a wide range of ecological concepts are covered, such as competition and predation, species conservation, parasitic infections, climate change, public attitudes about nature, reproduction and foraging ecology, unique morphology, habitat use, and the native or nonnative status of species. The stories focus on species of lampreys, gars, freshwater eels, pikes, minnows, suckers, catfishes, trouts, trout-perches, sculpins, sunfishes, and perches, a list that represents only about half of the fish families of the central Appalachians.

The chapters in this book are not comprehensive regarding old-school naturalists, all ecological concepts, or all fishes of the central Appalachians. There are many more ecological questions to ask and many more stories left to tell. Fish families of the central Appalachians not covered in this book include sturgeons, paddlefishes, bowfins, herrings, mudminnows, silversides, killifishes, livebearers, sticklebacks, striped bass, and drums. I hope to write a future companion volume including stories of the remaining families with emphasis on additional ecological concepts and other influential naturalists.

An understanding of nature is important to many people, and it is also important to the future of central Appalachian fishes. As more people learn about the ecology and diversity of our fishes, we improve our chances of protecting and conserving these species. I hope that the ecological narratives in this book inform and influence people's views about nature and naturalists and promote further interest on the nature of central Appalachian fishes.

Common and scientific names of the fishes found in this book,
listed in phylogenetic order by family

COMMON NAME	SCIENTIFIC NAME
Ohio Lamprey	*Ichthyomyzon bdellium*
Northern Brook Lamprey	*Ichthyomyzon fossor*
Mountain Brook Lamprey	*Ichthyomyzon greeleyi*
Silver Lamprey	*Ichthyomyzon unicuspis*
Least Brook Lamprey	*Lampetra aepyptera*
American Brook Lamprey	*Lethenteron appendix*
Longnose Gar	*Lepisosteus osseus*
American Eel	*Anguilla rostrata*
Northern Pike	*Esox lucius*
Muskellunge	*Esox masquinongy*
Chain Pickerel	*Esox niger*
Redfin Pickerel	*Esox americanus*
Stoneroller Minnow	*Campostoma anomalum*
Goldfish	*Carassius auratus*
Southern Redbelly Dace	*Chrosomus erythrogaster*
Mountain Redbelly Dace	*Chrosomus oreas*
Redside Dace	*Clinostomus elongatus*
Rosyside Dace	*Clinostomus funduloides*
Spotfin Shiner	*Cyprinella spiloptera*
Steelcolor Shiner	*Cyprinella whipplei*
Common Carp	*Cyprinus carpio*
Tonguetied Minnow	*Exoglossum laurae*
Cutlip Minnow	*Exoglossum maxillingua*
Striped Shiner	*Luxilus chrysocephalus*
Common Shiner	*Luxilus cornutus*
Allegheny Pearl Dace	*Margariscus margarita*
Hornyhead Chub	*Nocomis biguttatus*
Bluehead Chub	*Nocomis leptocephalus*

River Chub	*Nocomis micropogon*
Bigmouth Chub	*Nocomis platyrhynchus*
Bull Chub	*Nocomis raneyi*
Emerald Shiner	*Notropis atherinoides*
Spottail Shiner	*Notropis hudsonius*
Rosyface Shiner	*Notropis rubellus*
New River Shiner	*Notropis scabriceps*
Mimic Shiner	*Notropis volucellus*
Bluntnose Minnow	*Pimephales notatus*
Fathead Minnow	*Pimephales promelas*
Bullhead Minnow	*Pimephales vigilax*
Kanawha Minnow	*Phenacobius teretulus*
Longnose Dace	*Rhinichthys cataractae*
Creek Chub	*Semotilus atromaculatus*
Fallfish	*Semotilus corporalis*
Longnose Sucker	*Catostomus catostomus*
White Sucker	*Catostomus commersonii*
Northern Hogsucker	*Hypentelium nigricans*
Smallmouth Buffalo	*Ictiobus bubalus*
Bigmouth Buffalo	*Ictiobus cyprinellus*
Black Buffalo	*Ictiobus niger*
Golden Redhorse Sucker	*Moxostoma erythrurum*
Torrent Sucker	*Thoburnia rhothoeca*
Yellow Bullhead	*Ameiurus natalis*
Blue Catfish	*Ictalurus furcatus*
Channel Catfish	*Ictalurus punctatus*
Flathead Catfish	*Pylodictis olivaris*
Mountain Madtom	*Noturus eleutherus*
Stonecat Madtom	*Noturus flavus*
Tadpole Madtom	*Noturus gyrinus*
Margined Madtom	*Noturus insignis*
Brindled Madtom	*Noturus miurus*
Northern Madtom	*Noturus stigmosus*
Brook Trout	*Salvelinus fontinalis*
Trout-perch	*Percopsis omiscomaycus*
Banded Killifish	*Fundulus diaphanus*
Western Mosquitofish	*Gambusia affinis*

Bluestone Sculpin	*Cottus* sp. cf. *carolinae*
Buckeye Creek Cave Sculpin	*Cottus* sp. cf. *carolinae*
Checkered Sculpin	*Cottus* sp. cf. *cognatus*
Mottled Sculpin	*Cottus bairdii*
Blue Ridge Sculpin	*Cottus caeruleomentum*
Slimy Sculpin	*Cottus cognatus*
Potomac Sculpin	*Cottus girardi*
Kanawha Sculpin	*Cottus kanawhae*
White Perch	*Morone americana*
Rock Bass	*Ambloplites rupestris*
Green Sunfish	*Lepomis cyanellus*
Bluegill	*Lepomis macrochirus*
Smallmouth Bass	*Micropterus dolomieu*
Spotted Bass	*Micropterus punctulatus*
Largemouth Bass	*Micropterus salmoides*
Yellow Perch	*Perca flavescens*
Eastern Sand Darter	*Ammocrypta pellucida*
Crystal Darter	*Crystallaria asprella*
Diamond Darter	*Crystallaria cincotta*
Logperch	*Percina caprodes*
Appalachia Darter	*Percina gymnocephala*
Dusky Darter	*Percina sciera*
Greenside Darter	*Etheostoma blennioides*
Rainbow Darter	*Etheostoma caeruleum*
Fantail Darter	*Etheostoma flabellare*
Johnny Darter	*Etheostoma nigrum*
Kanawha Darter	*Etheostoma kanawhae*
Tessellated Darter	*Etheostoma olmstedi*
Candy Darter	*Etheostoma osburni*
Orangethroat Darter	*Etheostoma spectabile*
Tippecanoe Darter	*Etheostoma tippecanoe*
Variegate Darter	*Etheostoma variatum*

Responses by 37 individuals of Central Appalachia to an 1883 US Fish Commission questionnaire on the edibility of Common Carp. Responses are taken from Smiley[1] and arranged in alphabetical order by last name.

SUPERB.—We ate one, which was superb.—PETER BONDS, *Harrisonburg, Virginia*

GOOD.—I heard my son say he ate one, and that it was good.—DAVID BOWMAN, *Timberville, Virginia*

GOOD.—Have eaten one fried, and found it good.—SOLON M. BOWMAN, *Timberville, Virginia*

COULD NOT TELL.—We have fried and eaten two of the fish that we found in the grass with some hook holes through their mouths. Thieves had dropped them. We could not tell much about the quality.—JOHN B. BROWN, *Nashville, Ohio*

VERY FINE.—I ate two of them. I had them baked. Their edible qualities were very fine.—JAMES BUMGARDNER, SR., *Greenville, Virginia*

VERY GOOD.—The one I caught was fried, and was thought very good.—R. WELBY CARTER, *Upperville, Virginia*

DELICIOUS.—Only upon one occasion, and fried. It was delicious.—J. W. DOWNEY, M.D., *New Market, Maryland*

EQUAL TO BLACK BASS.—I think they are as good as black bass.—W. M. DOWNEY, *New Market, Maryland*

VERY PALATABLE.—Fried in bacon grease they are very palatable.—JOHN M. FERGUSON, *Alderson, West Virginia*

VERY FINE.—In my absence one of the oldest was caught with a hook. When landed the line parted, and the fish was injured so that it could not be returned to the pond. It was fried and pronounced by the family very fine.—JAS. G. FIELD, *Gordonsville, Virginia*

FRIED IN BUTTER AND LARD.—Of the very finest. One, which was rolled in wheat flour and fried in butter and lard. Their eatable qualities were of the very finest.—BENJAMIN L. GARBER, *Marietta, Pennsylvania*

SKIN, AND FRY OR BAKE THEM.—We first skin them, then thoroughly scald them and either fry or bake them.—O. A. GILMAN, *Paris, Kentucky*

VERY NICE.—We have cooked them two or three ways, and find that the larger ones are very nice.—O. A. GILMAN, *Paris, Kentucky*

GOOD.—We have eaten three fried. Their quality was good. I like them well.—J. B. HAGER, *Board Tree, West Virginia*

EXTRA.—We have caught and eaten some fried, and claim them to be extra in quality.—S. O. HAWKINS, *Bucks, Ohio*

VERY TOOTHSOME.—Yes; fried; and pronounced as very toothsome by all who had the pleasure of partaking of them.—JACOB G. HEILMAN, *Jonestown, Pennsylvania*

SOFT.—We ate two, fried in butter. They were pleasant tasted, but the flesh was most too tender.—J. W. HIGBEE, *Castle Shannon, Pennsylvania*

SUPERIOR TO ANY OTHER FISH THERE.—We have eaten two which were fried. They were delightful and have superior qualities over any other fish here.—W. M. L. HUDSON, *Luray, Virginia*

GOOD.—I have eaten but one, and was pleased with its flavor.—E. B. ISETT, *Spruce Creek, Pennsylvania*

GOOD AS HE WANTS.—I have eaten 2 of the first lot merely to try them. The quality vas as good as I want. They were only fried.—M. B. E. KLINE, *Broadway Depot, Virginia*

EXCELLENT.—In the summer of 1882, with hook and line, I caught three, one weighing 3 ½ pounds and the other two 2 ¼ and 2 ½ pounds, respectively. They were fried, and pronounced by the company to be "excellent."—JOSEPH LIGON, *Massie's Mills, Virginia*

EXCELLENT, PERHAPS SOFT.—I ate two fried. Their table qualities were excellent. The first one was a little too fat and the meat a little soft.—C. C. LOBINGIER, *Braddock, Pennsylvania*

SALTED AND FRIED BROWN: EXCELLENT.—We took one out of

pond No. 2, in May, 1883, weighing three pounds. When scaled, salted five hours, floured, and fried brown, it was of excellent flavor.—J. W. LONG, *Mount Morris, Pennsylvania*

VERY FINE, SWEET, AND RICH.—All report them very fine eating; very fat, sweet, rich, and toothsome when fried.—H. C. LOOSE, *Hagerstown, Maryland*

GOOD.—We ate one fried, and pronounced it good.—S. P. MCFALL, *Newton Falls, Ohio*

A LITTLE MUDDY TASTE.—I have eaten some fried in lard that was fat, very good except a little musty taste.—ANDREW MANN, *Forest Hill, West Virginia*

GOOD AS ANY.—We have baked them and they are as good as any fish we ever ate.—J. SHAW MARGERUM, *Washington, Pennsylvania*

VERY GOOD.—I had one fried, and considered it a very good fish. It weighed 2 ½ pounds.—W. J. PRICE, *Fincastle, Virginia*

OILY AND FINEST FISH HE EVER ATE.—We have eaten nine and given away three. We fried them like other fish. They contained nearly oil enough to cook themselves and were very fine—finest I ever ate.—HENRY PULSE, *Harrisonburg, Virginia*

GOOD.—Two. Fried in butter. Good.—LEWIS W. RUNNER, *Morgantown, West Virginia*

FRIED IN BUTTER: NICE.—We have eaten one; cleaned it in evening, salted it, and fried it in butter; thought it good and nice.—W. M. SADLER, *New Salem, West Virginia*

GOOD AND VERY GOOD.—We have eaten one. It was fried in lard, and was pronounced good and very good.—MICHAEL SHANK, *Harrisonburg, Virginia*

VERY GOOD.—Yes. In the winter of 1882–'83, the pond was drained by muskrats and the carp were killed. The older ones then weighed nearly 3 pounds. They were considered very good eating.—RUSH TAGGART, *Salem, Ohio*

EQUAL TO SHAD.—We ate one this morning. It was broiled. It was very good—something like shad.—W. M. THOMPSON JR., *Lemont, Pennsylvania*

CAUSE OF MUDDY TASTE.—I have eaten them both fried and broiled. I think the scale carp superior to the leather, but the quality of the carp depends upon their food. If left to care for themselves they will taste of the mud.—L. TRIPLETT JR., *Mount Jackson, Virginia*

PARTLY BOILED AND PARTLY BAKED: FIT FOR A KING.—I always instruct the cook to clean them nicely; then wrap the fish in a linen towel, have a large kettle of water boiling, coil the fish neatly in the kettle and boil fifteen minutes, then turn off the water, remove to a baking pan without marring and put in the oven, bake and then baste with butter gravy. A nice dressing could occupy the interior of the fish and the space around the sides. If properly done it makes a dish fit for a king or a hungry fisherman.—W. VAN ANTWERP, *Mount Sterling, Montgomery County, Kentucky*

BEST THEY EVER ATE.—We have eaten one which was fried. It was decidedly the best we ever ate. This was testified to by several.—JOHN C. WENGER, *Dayton, Virginia*

GOOD.—One small scale carp, accidentally killed in draining the pond, was fried as a pan-fish, eaten in my family, and pronounced good.—C. S. WHITE, *Romney, West Virginia*

TOO SOFT.—I do not think them a good pan fish, being too soft. They are good boiled.—ROBERT E. WITHERS, *Wytheville, Virginia*

BOILED LIKE ROCK: GOOD.—Yes; fried and boiled. The larger carp boiled and served as rock are served is palatable and good.—ROBERT E. WITHERS, *Wytheville, Virginia*

DELICIOUS.—Only once. They were fried in the usual way and were pronounced very palatable and delicious.—DANIEL WOLFE, *Fairplay, Maryland*

NEVER ATE BETTER FISH.—Have fried them, and never ate better fish.—W. M. O. YAGER, *Luray, Virginia*

Common names, scientific names, and date of description for 36 species of fishes from the Ohio River drainage discovered by Constantine Samuel Rafinesque, listed in phylogenetic order by family. Parentheses indicate a change in the genus name following the original species description.

Lake Sturgeon, *Acipenser fulvescens* Rafinesque, 1817

Shovelnose Sturgeon, *Scaphirhynchus platorynchus* (Rafinesque, 1820)

Shortnose Gar, *Lepisosteus platostomus* Rafinesque, 1820

Goldeye, *Hiodon alosoides* (Rafinesque, 1819)

Skipjack Herring, *Alosa chrysochloris* (Rafinesque, 1820)

Central Stoneroller, *Campostoma anomalum* (Rafinesque, 1820)

Southern Redbelly Dace, *Chrosomus erythrogaster* (Rafinesque, 1820)

Bigeye Chub, *Hybopsis amblops* (Rafinesque, 1820)

Striped Shiner, *Luxilus chrysocephalus* Rafinesque, 1820

Emerald Shiner, *Notropis atherinoides* Rafinesque, 1818

Bluntnose Minnow, *Pimephales notatus* (Rafinesque, 1820)

Fathead Minnow, *Pimephales promelas* Rafinesque, 1820

River Carpsucker, *Carpiodes carpio* (Rafinesque, 1820)

Highfin Carpsucker, *Carpiodes velifer* (Rafinesque, 1820)

Smallmouth Buffalo, *Ictiobus bubalus* (Rafinesque, 1818)

Black Buffalo, *Ictiobus niger* (Rafinesque, 1819)

Spotted Sucker, *Minytrema melanops* (Rafinesque, 1820)

Silver Redhorse, *Moxostoma anisurum* (Rafinesque, 1820)

Golden Redhorse, *Moxostoma erythrurum* (Rafinesque, 1818)

Black Bullhead, *Ameiurus melas* (Rafinesque, 1820)

Channel Catfish, *Ictalurus punctatus* (Rafinesque, 1818)

Stonecat, *Noturus flavus* Rafinesque, 1818

Flathead Catfish, *Pylodictis olivaris* (Rafinesque, 1818)

Blackstripe Topminnow, *Fundulus notatus* (Rafinesque, 1820)

White Bass, *Morone chrysops* (Rafinesque, 1820)

Rock Bass, *Ambloplites rupestris* (Rafinesque, 1817)

Green Sunfish, *Lepomis cyanellus* Rafinesque, 1819

Bluegill, *Lepomis macrochirus* Rafinesque, 1819

Longear Sunfish, *Lepomis megalotis* (Rafinesque, 1820)

Spotted Bass, *Micropterus punctulatus* (Rafinesque, 1819)

White Crappie, *Pomoxis annularis* Rafinesque, 1818

Greenside Darter, *Etheostoma blennioides* Rafinesque, 1819

Fantail Darter, *Etheostoma flabellare* Rafinesque, 1819

Johnny Darter, *Etheostoma nigrum* Rafinesque, 1820

Logperch, *Percina caprodes* (Rafinesque, 1818)

Freshwater Drum, *Aplodinotus grunniens* (Rafinesque, 1819)

GLOSSARY

acid mine drainage—The acidic water flowing from a mining site, often resulting from oxidation of iron sulfide.

adipose fin—A small, fleshy, dorsally located fin of some fishes that occurs between the dorsal fin and the caudal fin.

algae—A diverse group of simple photosynthetic organisms that includes diatoms, green algae, and cyanobacteria.

algivore—An animal that eats algae.

aluminum—A chemical element and the most abundant metal of the earth's crust.

ammocoete—The larval life stage of a lamprey.

anadromous—A fish that migrates from the sea to spawn in freshwater rivers.

anal fin—A ventral fin of a fish located posterior of the anus.

annelid—A diverse group of segmented worms.

antimicrobial—An agent that kills or stops the growth of microorganisms.

archaeology—The study of material objects of people from past history and civilizations.

arthropod—An invertebrate animal with external skeleton (exoskeleton), body segmentation, and paired, jointed appendages.

articular physiology—The study of the function and movements of bone joints.

autotroph—An organism at the base of the food chain, often referred to as a primary producer, that converts simple inorganic matter to complex and nutritional organic compounds.

baby schema—A set of physical features of an infant that people find cute and attractive, including a big head and large eyes.

background matching—A type of animal concealment where the pattern of body pigmentation is similar to that of the surrounding environment.

bait-bucket introduction—An act, often by an angler, where live fish are released from a bait bucket into a water body outside of their natural distribution range.

barbel—A fleshy protuberance located near the mouth of individuals of certain fish species.

basicaudal—A region of a fish's body located at the base of the tail fin.

benthic—On or near the bottom of a stream or lake.

Bernoulli principle—A reduction in pressure associated with increased velocity.

biodiversity—The variety of life on earth, or of life within a specified geographic area.

black spot disease—A parasitic infection where a fish's body has black spots resulting from encysted trematodes.

blue rooter—A colloquial name for the Black Buffalo fish.

bottom feeder—A fish that eats on or near the bottom of a lake or stream.

bottom-up effect—An effect where changes in a lower level of a food chain result in changes to a higher level.

boulder—A particle or grain size of rocks with the longest axis larger than 10.1 inches (256 mm).

boundary layer effect—A friction-induced phenomenon where water velocities slow down next to a submerged surface.

breeding territory—An area that is used and defended by an animal for reproductive purposes.

breeding tubercle—A small keratinized projection, also called a pearl organ or nuptial tubercle, that develops on the body or fins of some fishes just before or during their breeding season.

bugle-mouth—A colloquial name for the Common Carp.

cabinet of curiosities—A room or area with a large number of interesting items.

caddisfly—An aquatic insect of the order Trichoptera.

calcium—A chemical element classified as an alkaline earth metal.

captive propagation—The rearing of fish in captivity.

carbonate rock—A class of sedimentary calcium-based rocks including limestone and dolomite.

cardiovascular—A system of the body that circulates blood and nutrients.

carnivore—An animal that eats meat.

cascade—A series of small waterfalls on a steep gradient section of a stream.

catadromous—A fish that migrates from a freshwater river to spawn in the sea.

caterpillar—The larval stage of a moth or butterfly.

Catostomidae—The family name of a taxonomic fish group known as suckers.

caudal fin—The tail fin of a fish.

centripetal positioning—Moving toward a center.

cercaria—A parasitic trematode's free-swimming larval stage that passes from its intermediate host to another host (plural cercariae).

charismatic species—A species viewed as important based on certain physical characteristics, such as large body size, aesthetic attractiveness, or bright coloration.

cobble—A particle or grain size of rocks with the longest axis ranging from 2.5 to 10.1 inches (64–256 mm).

colony nesting—The clustering of nests within a common area by a group of individuals.

commensalism—A relationship between species in which only one species benefits.

conical—Cone-shaped. In this book, a word used to describe the snout shape of a Logperch.

conservationist—A person who advocates for the preservation of the diversity of life.

countershading—A type of concealment that occurs in animals with a dark dorsum and a light ventrum. Downwelling light illuminates the dark dorsum, and a shadow effect darkens the light ventrum.

crypsis—An animal's ability to not be seen or detected by other animals.

ctenoid scale—A thin fish scale with growth rings (annuli) and small spines (ctenii) on its posterior surface.

cuckholder male—A reproductively active fish, such as a female mimic or sneaker male, that cuckholds an older dominant male during the spawning process.

cyanobacteria—A type of bacteria often called blue-green algae.

cycloid scale—A thin smooth-edged fish scale with growth rings (annuli) that lacks small spines on its posterior surface.

darter—A group of fishes classified within the family Percidae and represented by four genera: *Ammocrypta, Crystallaria, Etheostoma,* and *Percina.*

dear enemy effect—A phenomenon in which aggressive behavior lessens between two neighboring territory-holding animals after the two individuals become accustomed to one another.

deciduous hardwood trees—Species of trees that seasonally lose their leaves.

deflection hypothesis—A term used in predator-prey interactions where pigmentation or movement of the prey misleads a predator to attack a less vulnerable location of the prey's body.

depauperate—An ecosystem or geographic area with a small number of species.

detritus—Dead organic matter.

diatom—A single-celled algae structured with silica.

dilatancy—A phenomenon where a granular substance becomes more viscous when under pressure.

dilution effect—A predation-risk concept where an increase in the number of individuals in a group reduces an individual's probability of being eaten by a predator.

disruptive coloration—A type of concealment where bold markings disrupt or break up an animal's body outline.

dorsal fin—A fin on the back of a fish.

dorsal saddle—A narrow to broad stripe of pigmentation across the back of a fish.

dorsum—The upper or back area of a fish's body.

drift feeding—A foraging tactic where downstream water currents carry food items to a stationary fish.

ecosystem engineer—An animal, also known as a keystone modifier, that alters a habitat, and the changed habitat is then used by other species.

eddy—An area in a stream with noticeably slower water velocities than surrounding areas, often just downstream of a velocity barrier, such as a partially submerged rock or log.

eelway—A fish ladder that is designed to allow eels to pass upstream.

egg mimic—An anatomical structure of an animal that has a similar appearance to that of an egg, or an egg of one species that has a similar appearance to that of another species.

elver—An early developmental stage of anguillid eels following the glass eel stage where the body changes from being transparent to pigmented.

endemic species—A species that is native to a drainage area.

ethologist—A scientist involved with the comparative study of animal behavior.

evapotranspiration—The transfer of land-based water to the atmosphere by evaporation of moisture from the soil and other surfaces and by plant transpiration.

excretion—The process where a living organism expels waste products into the environment, including nutrients such as nitrogen and phosphorus. Examples include urinary and fecal excretion.

extirpation—The loss of a population of a species.

extinction—The loss of a species.

exotic species—A species that has been moved from its natural distribution range into another geographic area. Often used as a synonym for "introduced species," "nonindigenous species," and "nonnative species," but sometimes it describes an introduced species whose native distribution range occurs in another country or another continent.

eyespot—A blotch of pigment on an animal's body that mimics an eye.

Fallingwater House—A summer home built during 1936–37 on top of a waterfall on Bear Run in the Laurel Highlands of central Pennsylvania. The home was designed by Frank Lloyd Wright and is registered as a National Historic Landmark.

fast-start foraging—A predator's high-powered burst of acceleration toward its prey from a resting or near-resting position.

fecundity—A measure of the ability to produce offspring.

female mimic—A male fish that behaves and looks like a female and cuckolds an older dominant male during the spawning process.

filial cannibalism—The occurrence of an animal eating all or some of its offspring.

filter feeding—The act of foraging by an animal with a specialized filtering structure that strains suspended small particles or small living organisms from the water column.

fin spine—An element of a fish fin that is unsegmented and usually stiff with a sharp point. Some species, such as sculpins, have soft fin spines.

fin standing—A behavior in which a fish uses a combination of pelvic, pectoral, anal, or caudal fins to prop its body off the stream or lake bottom.

fitness—A term that relates to an individual's ability to survive and produce offspring.

flagship species—A species that serves as a symbol for promoting conservation awareness.

food chain—Organisms that are serially linked by the transfer of energy and nutrients. For example, autotrophs are on the bottom end of the food chain and carnivores are at the top end.

food web—An intricate pattern of a series of interconnected food chains.

foraging area—An area where an animal goes to find food.

forest pathogen—An organism, often of microscopic size, that can reduce the health of trees or cause tree death.

fossil pollen—Pollen grains of plants that have been preserved in lake sediments and peatlands and are used by scientists to reconstruct past climates.

foundation species—A species that influences biodiversity and ecosystem function.

frictional drag—A downstream-directed force acting on a submerged body resulting from friction between the body's surface and adjacent moving fluid.

ganoid scale—A type of fish scale of certain fishes, such as gars and sturgeons, usually with a rhomboid shape and with an enamel-like covering known as ganoine.

gape-limited—An animal that is restricted as to how far it can open its mouth. The term is used in reference to a predator that cannot bite off or tear off a piece of its prey, but must eat its prey whole.

geographic distribution range—An area where a species or taxon lives, often determined by occurrence records.

glacier—A large ice mass that moves slowly over land.

glass eel—An early developmental stage of anguillid eels where the body is transparent.

glide—A transitional stream habitat with shallow depths, moderate water velocities, and smooth water surface.

Goldilocks effect—When there is a perceived choice of liberal, moderate, and conservative conditions and the moderate condition is preferred (based on the fairy tale "Goldilocks and the Three Bears").

gourdhead—A colloquial name of the Bigmouth Buffalo fish.

gravel—A particle or grain size of rocks with the longest axis ranging from 0.079 to 2.5 inches (2–64 mm).

habitat fragmentation—The loss of parts of a contiguous tract of habitat, resulting in smaller isolated patches of habitat.

habitat shift—A change of habitat use by an animal.

hanging culvert—A pipe that allows a stream to pass underneath a roadway where its downstream outlet is elevated above the water's surface.

headwater stream—A stream located in the upper reaches of a drainage network.

Helmholtz square illusion—An illusion that occurs when viewing two adjacent equally sized squares, one with vertical stripes and the other with horizontal stripes. The square with vertical stripes appears to be wider, and the square with horizontal stripes appears to be taller and narrower. The concept is named in honor of Hermann von Helmholtz, a German physician and physicist.

Hemlock Woolly Adelgid—A small aphid-like insect from Japan that has been introduced into the United States, where it is considered a major threat to Hemlock species.

herbivore—An animal that eats plant matter or algae. Animals that eat algae are also called algivores.

hermaphroditic—An organism with both female and male sexes.

herpetology—The study of reptiles and amphibians.

Holocene—The current geologic time period, which began about 11,700 years ago.

hornyhead—A colloquial name for fishes that develop head tubercles.

host—An individual that harbors a parasite.

hot spring—A spring with water temperature that exceeds 100°F.

hybridization—The production of hybrid offspring that results from the mating of individuals of two species.

hydraulic fracturing—A process used for oil and gas wells, also known as fracking, where deep rock formations are fractured by fluid injected under high pressure into a well bore.

hydrofoil—An asymmetric wing-like structure that produces lift when it moves through water or when it is stationary in moving water. The lift results from the Bernoulli effect, where increased water velocity causes less pressure on top of the asymmetric hydrofoil.

hydrogen ion concentration—A measure of hydrogen ions in solution that is often converted to the pH logarithmic scale that ranges from 0 to 14.

ichthyology—The study of fishes.

inferior mouth position—A mouth that is positioned ventrally on the head of a fish.

interstitial space—A cavity between closely positioned objects.

intermediate host—An individual that harbors an early life stage of a parasite.

introduced species—A species that has been moved from its natural distribution range into another geographic area. Commonly used as a synonym for "nonindigenous species" and "nonnative species."

introgression—The transfer of genes from one species to another that results from subsequent mating between a hybrid individual and an individual from one of its parental species.

invertebrate—An animal without a backbone.

invertivore—An animal that eats invertebrates.

larvae—A name used for individuals of some animals during an immature or early life stage (singular larva).

Joe-Pye weed—A common name for a plant in the genus *Eutrochium*.

Karpfen—German name for the Common Carp.

karst area—A geographic area underlain by limestone bedrock and often including sinkholes and underground caverns.

keratin—A structural protein found in vertebrates.

kinetic theory—A theory that explains the motion and interaction properties of gases.

laminar flow—Movement of a fluid in smooth parallel layers.

lamprey—An eel-shaped jawless fish of the family Petromyzontidae.

Latin binomial—The two-word format for a scientific name of a species, where the first word is the genus and the second word is the specific epithet.

lie-in-wait—A foraging strategy where an animal waits on the arrival of a food item.

Little Ice Age—A cold period lasting from about 530 years ago to the early 1900s.

leptocephalus—A small ribbonlike fish larva.

macroinvertebrates—Small animals without backbones that can be observed without the use of a microscope.

mammalogy—The study of mammals.

mayfly—An aquatic insect of the order Ephemeroptera.

median fins—Fins that are located along the midplane of a fish's body. The midplane, also called a median plane or midsagittal plan, divides the fish's body in left and right sides. Examples include the dorsal, adipose, caudal, and anal fins.

Medieval Warm Period—A warm period lasting from approximately 1,120 to 530 years ago, occurring just before the Little Ice Age.

metacercaria—A parasitic trematode's larval stage that has transformed from a cercarial larvae and encysts in the tissue of the host species (plural metacercariae).

metamorphose—The act of changing body form or shape that is experienced by some animals as a developmental process.

meteorology—The study of climate and weather.

midge—A general name for various species of tiny flies, most of which have an aquatic larval stage.

miller's thumb—A colloquial fish name for a sculpin of the genus *Cottus*.

mimicry—The close resemblance of one organism to another.

minnow—A common name for fishes in the family Leuciscidae (formerly family Cyprinidae).

miracidia—A parasitic trematode's free-swimming larval stage that infects its first host species (plural miracidium).

mixed-species shoal—A social gathering of more than one fish species.

morph—A variant body form of an animal.

morphology—The form and structure of external and internal body parts.

mortality rate—A measure of the number of deaths in a population.

mouth gape—The widest position that an animal can open its mouth.

mud cat—A colloquial name used for species of North American catfish, including the Flathead Catfish.

muddler minnow—An artificial streamer tied by a fly angler that mimics a sculpin.

nasty neighbor effect—A within-species phenomenon where a territory-holding animal exhibits more aggressive behavior toward its neighbor than it does toward unfamiliar individuals.

native fish—A fish that occurs in a drainage area that is within its natural distribution range.

naturalist—A person with a keen interest in the details of nature.

nest associates—Individuals of a species or a group of species that occur within or near the spawning nest of another species.

nitrate—A chemical structure of nitrogen with an ion composed of one nitrogen atom and three atoms of oxygen.

nitrogen cycling—The cycle or circulation of nitrogen, during which it is converted into multiple chemical forms.

nocturnal—An animal that is active at nighttime.

nonindigenous species—A species that has been moved from its natural distribution range into another geographic area. Commonly used as a synonym for "introduced species" and "nonnative species."

nonnative species—A species that has been moved from its natural distribution range into another geographic area. Commonly used as a synonym for "introduced species" and "nonindigenous species."

nuclear-follower relationship—A stream-bottom foraging interaction in which food items dislodged by one or more individuals of a species drift downstream and are eaten by one or more individuals of another species. The fish that dislodge the food items represent the nuclear species, and the fish foraging downstream represent the follower species.

nuptial crest—A hump that develops on top of the head of some fishes just before or during the breeding season.

nuptial male—A male that is ready to spawn.

nuptial tubercle—A small keratinized projection, also called a breeding tubercle or pearl organ, that develops on the body or fins of some fishes just before or during their breeding season.

nutrients—Chemicals or elements necessary for the survival of organisms, which cycle back and forth between living organisms and the physical environment.

nutrient spiraling—The movement of nutrients in a stream, involving temporary delays in downstream transport.

old-school naturalist—A person from an earlier time period with a keen interest in nature and a propensity to record detailed field notes from their observations on the natural world.

operculum—A series of fish bones that provides facial structure and protective cover of gills.

organic architecture—Design and construction of buildings or structures that are balanced or melded with the natural surroundings.

ornithologist—A person with a keen interest in birds.

palindromic swimming—The ability to swim backward as well as swim forward.

papillose lips—Fish lips with a pimpled surface texture.

parasitology—The study of parasites.

parental male—A large male fish that demonstrates dominance over smaller males of the same species during breeding periods.

pearl organ—A small keratinized projection, also called a breeding tubercle or nuptial tubercle, that develops on the body or fins of some fishes just before or during their breeding season.

pectoral fin—One of a pair of laterally or ventrolaterally positioned fins on a fish located just posterior of the head.

pelagic—Occurring in the water column of a stream or lake.

pelvic fin—One of a pair of ventrally located fins on a fish.

Percidae—The perch family of fishes.

Percopsidae—The family name for Trout-perch.

periphyton—The slippery organic covering on the surface of stream substrates.

piscivore—An animal that eats fish.

pit-ridge—The name of the nest type that is built by a breeding male Creek Chub.

plankton—A diverse group of small or microscopic aquatic organisms that are generally drifting or suspended in the water column.

Pleistocene—A geologic time period involving repeated glaciations that spanned from about 2,580,000 to 11,700 years ago.

plicate lips—Fish lips with a surface texture of parallel grooves.

plunge pool—A water-filled depression in a stream that forms at the base of a waterfall or at the base of a partially submerged structure.

polymath—A person with a wide range of knowledge.

pool—A habitat type of flowing streams with relatively deep, slow-moving water.

population—A group of individuals of the same species that occur within a certain geographic area.

precolonial Holocene—A geologic time period ranging from about 11,700 years ago to the time period of colonial occupation of a region.

predator—An animal that eats other animals.

predator inspection—The act of one or more individuals of a prey species that leave a social group to approach and get a closer look at a potential predator.

predation risk—The likelihood that an individual of a prey species will be eaten by a predator.

pressure drag—A downstream-directed force acting on the surface of a submerged body that results when the boundary layer separates from the body's surface, altering the pressure distribution over that body's surface and causing increased pressure in the direction of the moving fluid.

prey—An animal that is eaten by a predator.

primary production—The synthesis of complex organic compounds by autotrophs.

prisoner's dilemma—A game between two individuals where each player either benefits or is penalized based on cooperative or uncooperative behavior.

profile drag—A downstream-directed force acting on a submerged body resulting from a combination of both pressure and frictional drag.

protozoa—A taxonomic group of single-celled microscopic animals.

protrusible mouth—A fish mouth that can extend outward.

psammophilous—A term that refers to a sand-dwelling animal.

range reduction—The shrinkage of the geographic distribution range of a species.

razorback—A colloquial name of the Smallmouth Buffalo fish.

red-eye—A colloquial name for the Rock Bass.

refuge area—An area where an animal goes to find safety from a potential threat.

renaissance man—A man with many talents and a wide range of knowledge.

Reynolds number—A dimensionless number attributed to Osborne Reynolds, a British engineer and physicist. The number is calculated as the ratio of inertial force to viscous force of a moving fluid. The inertial force is associated with the fluid's momentum, where momentum is influenced by the density and velocity of the fluid. The viscous force results from friction between fluid layers that causes resistance to fluid flow.

riffle—A habitat type of flowing streams with a gradient, relatively shallow depths, fast water velocities, and a turbulent water surface.

riparian—A word derived from the Latin word *ripa*, meaning riverbank, referring to the interface between stream and land.

rotational feeding—A method of feeding where an anguillid eel spins its body after clamping its jaws on the flesh of a prey species, thus tearing off a small piece of flesh for consumption.

run—A habitat type of flowing streams with moderate depths, moderate water velocities, and a relatively smooth water surface.

Salmonidae—The family name for charr, salmon, and trout.

sand—A particle or grain size with a diameter ranging from 0.0025 to 0.0790 inches (0.0635–2.0 mm).

Sargasso Sea—An area of the Atlantic Ocean that spans Bermuda to the Azores.

schooling—A social gathering of fish or fishes with synchronized swimming and body positions oriented in the same direction.

selfish herd—A predation-risk theory that states the individuals within the center of a group or eggs within the center of a nest are safer than those positioned at or near the periphery.

sexual selection—A mode of natural selection that arises from differential reproductive success within a population where an individual's preferences for certain characteristics of the opposite sex influence mate choice.

sharptooth McGraw—A colloquial name for the Northern Pike.

Shay locomotive—A geared steam locomotive attributed to the inventor Ephraim Shay.

shoaling—A social gathering of closely positioned fish or fishes where body positions of many individuals are oriented in different directions.

shoalmate—A fish that is a member of a social group of individuals.

sinkhole—A depression in the ground caused by dissolution of carbonate rocks in a karst landscape.

silt—A particle or grain size with a diameter ranging from 0.00015 to 0.00250 inches (0.00381–0.06350 mm).

silver eel—The adult stage of anguillid eels.

smoke screen—A strategy used in military maneuvers where the cover of smoke or fog provides concealment.

smoke-screen foraging—A foraging strategy in which a bottom-feeding aquatic predator conceals itself from prey by creating a cloud or plume of suspended sediments. The sediment cloud provides the predator with a stealthy approach and may calm the prey.

sneaker male—An early maturing sneaky male fish that cuckholds an older dominant male during the spawning process.

spatial frequency—A term used in mathematics, physics, and engineering as a measure of how often a continuous wave form repeats per unit of distance. In this book, the term is used in reference to the repeating pigment bars on the side of a Logperch's body.

spawning—An act of reproduction in fishes, typically involving the female's release of unfertilized eggs and the male's release of milt.

spawning area—An area where fish reproduce.

spawning clasp—A reproductive behavior of some fishes where a male keeps a female's body close to his by using fins, tubercles, or body curvature.

spawning mound—An aggregation of gravels used as a nest for spawning.

species authority—The person recognized for describing a new species.

specific epithet—The second word in a Latin binomial.

splash dam—A wooden dam built for the purpose of raising water levels in a stream to aid the downstream transport of logs to sawmills.

spotted cat—A colloquial name used for juvenile Flathead Catfish in Central Appalachia.

S-start foraging—An S-shaped body position of some predatory fishes during the initial stages of attacking their prey.

station-holding—A fish's ability to maintain its position in a flowing stream.

stickleback—A common name for a fish of the family Gasterosteidae, named for the prominent spines of the dorsal fin.

stonefly—An aquatic insect of the order Plecoptera.

streamlined body—A shape or profile of a body that experiences reduced drag forces when submerged in a moving fluid.

stream sedimentation—The deposition of small-sized particles, such as silt, onto the bottom of a stream.

subwatershed—A drainage area nested within a larger watershed area, also referred to as subdrainage, subbasin, and subcatchment.

sucker—A fish in the family Catostomidae.

swim bladder—An internal gas-filled sac that some fishes use to control their buoyancy.

taxonomy—The science of naming organisms and classifying them into groups.

taxonomic nomenclature—A formal system of naming groups of organisms.

terrestrial invertebrates—Nonaquatic animals without backbones.

Thayer's law—Concealment by countershading.

tit-for-tat—A strategy in game theory where an individual initially cooperates with its opponent and then mimics the opponent's previous action during its subsequent moves.

top-down effect—An effect where changes in a higher level of a food chain result in changes to a lower level.

transpiration—A process where moisture moves upward inside of a plant, evaporating from stems and leaves.

trematode—A parasite in the order Trematoda.

trophic cascade—A directional change across levels of a food chain.

turbidity—A measure of the transparency loss of water caused by suspended particles.

turbulence—A term used to describe chaotic or erratic motion of wind or water.

turbulent flow—Chaotic movement of a fluid.

urbanization—The loss or reduction of a rural landscape owing to the design and development of cities or towns.

venomous—An organism that produces venom.

ventrum—The underside area of a fish's body.

viscosity—A measure of the thickness of a fluid. Water has a lower viscosity than syrup.

viscous sublayer—A friction-induced thin layer of fluid with a velocity near zero that forms on the surface of a submerged object in a moving fluid. Often depicted as a sublayer of the boundary layer.

warm spring—A spring with water temperatures of at least 9°F higher than mean annual air temperatures.

water velocity—The speed at which water is moving in a stream or river.

weir dam—A wall of stacked rocks or wooden structures crossing a river in a V shape, which funnels fishes into an enclosure.

yellow eel—A developmental stage of anguillid eels following the elver stage where body pigment is often yellowish to olive color.

NOTES

CHAPTER 1: LAMPREY ENLIGHTENED

1. Charles C. Abbott, *Notes of the Night and Other Outdoor Sketches* (New York: Century, 1896), 1.
2. Harold N. Segall, William H. Helfland, and Lloyd G. Stevenson, "Historic Marker for Charles Conrad Abbott Farm," *Journal of the History of Medicine and Allied Sciences* 22 (1967): 418–19.
3. Charles C. Abbott, "Descriptions of New Species of American Fresh-Water Fishes," *Proceedings of the Academy of Natural Sciences of Philadelphia* 12 (1860): 325.
4. Charles C. Abbott, *The Rambles of an Idler* (Philadelphia: George W. Jacobs, 1906), 253.
5. Ian C. Potter, Howard S. Gill, and Claude B. Renaud, "Petromyzontidae: Lampreys," in *Freshwater Fishes of North America*, ed. Melvin L. Warren Jr. and Brooks M. Burr (Baltimore: Johns Hopkins University Press, 2014), 105–39.
6. F. William H. Beamish and Jo-Anne Jebbink, "Abundance of Lamprey Larvae and Physical Habitat," *Environmental Biology of Fishes* 39 (1994): 209–14.
7. Ian C. Potter, "Ecology of Larval and Metamorphosing Lampreys," *Canadian Journal of Fisheries and Aquatic Sciences* 37 (1980): 1641–57.
8. M. W. Hardisty and Ian C. Potter, *The Biology of Lampreys,* vol. 1 (London: Academic, 1971), 1–423.
9. Milton B. Trautman, *The Fishes of Ohio* (Columbus: Ohio State University Press, 1981), 162–63.
10. Margaret F. Docker, "A Review of the Evolution of Nonparasitism in Lampreys and an Update of the Paired Species Concept," in *Biology, Management, and Conservation of Lampreys in North America,* ed. Larry R. Brown, Shawn D. Chase, Matthew G. Mesa, Richard J. Beamish, and Peter B. Moyle (Bethesda, MD: American Fisheries Society, 2009), 71–114; Margaret F. Docker, Nicholas E. Mandrak, and Daniel D. Heath, "Contemporary Gene Flow between 'Paired' Silver (*Ichthyomyzon unicuspis*) and Northern Brook (*I. fossor*) Lampreys: Implications for Conservation," *Conservation Genetics* 13 (2012): 823–35.
11. H. A. Surface, "Removal of Lampreys from the Interior Waters of New York," in *Fourth Annual Report, Commission of Fisheries, Game and Forests of the State of New York* (Albany, NY: Wynkoop Hallenbeck Crawford, 1899), 191–245.
12. Philip A. Cochran, "Predation on Lampreys," *American Fisheries Society Symposium* 72 (2009): 139–51.
13. Scott H. Ensign and Martin W. Doyle, "Nutrient Spiraling in Streams and River Networks," *Journal of Geophysical Research* 111 (2006): G04009, https://doi.org/10.1029/2005JG000114.
14. Bashford Dean and Francis B. Sumner, "Notes on the Spawning Habitats of the Brook Lamprey (*Petromyzon wilderi*)," *Transactions of the New York Academy of Sciences* 16 (1897): 321–24.

15. Philip A. Cochran, Devin D. Bloom, and Richard J. Wagner, "Alternative Reproductive Behaviors in Lampreys and Their Significance," *Journal of Freshwater Ecology* 23 (2008): 437–44.

16. Herbert F. Seversmith, "Distribution, Morphology, and Life History of *Lampetra aepyptera*, a Brook Lamprey, in Maryland," *Copeia* 1953 (1953): 225–32.

17. Cochran, "Predation on Lampreys," 139–51.

18. Margaret F. Docker and F. William H. Beamish, "Growth, Fecundity, and Egg Size of Least Brook Lamprey, *Lampetra aepyptera*," *Environmental Biology of Fishes* 31 (1991): 219–27.

19. Martin W. Hardisty, "The Growth of Larval Lampreys," *Journal of Animal Ecology* 30 (1961): 357–71.

20. Seversmith, "Life History of *Lampetra aepyptera*," 225–32.

21. Dustin M. Smith, "Habitat Selection and Predation Risk in Larval Lampreys" (MS thesis, West Virginia University, 2009), 1–51.

22. Dustin M. Smith, Stuart A. Welsh, and Philip J. Turk, "Selection and Preference of Benthic Habitat by Small and Large Ammocoetes of the Least Brook Lamprey (*Lampetra aepyptera*)," *Environmental Biology of Fishes* 91 (2011): 421–28.

23. Dustin M. Smith, Stuart A. Welsh, and Philip J. Turk, "Available Benthic Habitat Type May Influence Predation Risk in Larval Lampreys," *Ecology of Freshwater Fish* 21 (2012): 160–63.

24. Charles C. Abbott, *In Nature's Realm* (Trenton: Albert Brandt, 1900), 39.

25. Charles C. Abbott, *Outings at Odd Times* (New York: D. Appleton, 1890), 209.

CHAPTER 2: ILL-REGARDED

1. Aldo Leopold, *A Sand County Almanac: And Sketches Here and There* (New York: Oxford University Press, 1949), 129–32.

2. Carl Linnaeus, *Systema naturae per regna tria naturae, secundum classes, ordines, genera, species, cum characteribus, differentiis, synonymis, locis,* 10th ed., vol. 1 (Stockholm: Laurentii Salvii, 1758), 313.

3. Bernard G. Lacépède, *Histoire naturelle des poissons, tome cinquième* (Paris: Plassan, 1803), 331; translated from French.

4. Louis Agassiz, *Recherches sur les poissons fossiles* (Neuchatel Suisse: Imprimerie de Petitpierre, 1833–44); translated from French.

5. Agassiz, 74.

6. William C. Williamson, "On the Microscopic Structure of the Scales and Dermal Teeth of Some Ganoid and Placoid Fish," *Philosophical Transactions of the Royal Society of London* 139 (1849): 473.

7. William L. Pflieger, *Fishes of Missouri* (Missouri Department of Conservation, 1975), 66.

8. Robert E. Jenkins and Noel M. Burkhead, *Freshwater Fishes of Virginia* (Bethesda, MD: American Fisheries Society, 1994), 198.

9. Harry Van Meter, "Gigging the Long-Nosed Gar," *West Virginia Conservation* (May 1956): 6–7.

10. Stephen A. Forbes and Robert E. Richardson, *The Fishes of Illinois,* 2nd ed. (Springfield: Illinois State Journal Co., State Printers, 1920), 41.

11. Forbes and Richardson, 32.
12. Forbes and Richardson, 33.
13. Alexander Agassiz, "The Development of Lepidosteus, Part I," *Proceedings of the American Academy of Arts and Sciences* 14 (1878): 65–76.
14. Herbert T. Boschung and Richard L. Mayden, *Fishes of Alabama* (Washington, DC: Smithsonian Books, 2004), 114.
15. Ancil D. Holloway, "Notes on the Life History and Management of the Shortnose and Longnose Gars in Florida Waters," *Journal of Wildlife Management* 18 (1954): 438–49.
16. Holloway, 445.
17. L. Hussakof, "Fishes Swallowed by Gar Pike," *Copeia* 1914, no. 11 (1914): 2.
18. Karl F. Lagler and Frances V. Hubbs, "Food of the Long-Nosed Gar (*Lepisosteus osseus oxyurus*) and the Bowfin (*Amia calva*) in Southern Michigan," *Copeia* 1940 (1940): 239–41.
19. Jack D. Tyler, "Food Habits, Sex Ratios, and Size of Longnose Gar in Southwestern Oklahoma," *Proceedings of the Oklahoma Academy of Science* 74 (1994): 41–42.
20. William K. Gregory, "Memorial of Bashford Dean, 1867–1928," in *The Bashford Dean Memorial Volume: Archaic Fishes, Part 1*, ed. E. W. Gudger (New York: American Museum of Natural History, 1930), 1–42.
21. Bashford Dean, "The Early Development of Gar-Pike and Sturgeon," *Journal of Morphology* 11 (1895): 1–62.
22. Dean, 4–5.
23. Leopold, *Sand County Almanac*, 129–32.
24. Henry W. Robison and Thomas M. Buchanan, *Fishes of Arkansas* (Fayetteville: University of Arkansas Press, 1988).
25. William H. Clay, *The Fishes of Kentucky* (Frankfort: Kentucky Department of Fish and Wildlife Resources, 1975), 57.
26. Pflieger, *Fishes of Missouri*, 66.
27. Alfred C. Weed, *The Alligator Gar* (Chicago: Field Museum of Natural History, 1923), 10.
28. Kelly A. George, Kristina M. Slagle, Robyn S. Wilson, Steven J. Moeller, and Jeremy T. Bruskotter, "Changes in Attitudes toward Animals in the United States from 1978 to 2014," *Biological Conservation* 201 (2016): 237–42.
29. Dennis L. Scarnecchia, "A Reappraisal of Gars and Bowfins in Fishery Management," *Fisheries* 17 (1992): 6–12.

CHAPTER 3: EEL WAYS

1. Carl H. Strandberg and Ray Tomlinson, "Photoarchaeological Analysis of Potomac River Fish Traps," *American Antiquity* 34 (1969): 312–19.
2. James J. Kirkwood, *Waterway to the West* (Eastern National Park and Monument Association, 1963), 15.
3. William E. Davies, *The Geology and Engineering Structures of the Chesapeake and Ohio Canal; An Engineering Geologist's Descriptions and Drawings* (Glen Echo, MD: C&O Canal Association, 1999), xiv.
4. Emory L. Kemp, *The Great Kanawha Navigation* (Pittsburgh: University of Pittsburgh Press, 2000).

5. Alex Haro, William Richkus, Kevin Whalen, Alex Hoar, W.-Dieter Busch, Sandra Lary, Tim Brush, and Douglas Dixon, "Population Decline of the American Eel: Implications for Research and Management," *Fisheries* 25 (2000): 7–16; Douglas A. Dixon, ed., *Biology, Management, and Protection of Catadromous Eels, Symposium 33* (Bethesda, MD: American Fisheries Society, 2003); John M. Casselman and David K. Cairns, eds., *Eels at the Edge: Science, Status, and Conservation Concerns; Proceedings of the 2003 International Eel Symposium* (Bethesda, MD: American Fisheries Society, 2009).

6. Stephen D. Hammond and Stuart A. Welsh, "Seasonal Movements of Large Yellow American Eels Downstream of a Hydroelectric Dam, Shenandoah River, West Virginia," in *Eels at the Edge: Science, Status, and Conservation Concerns, Proceedings of the 2003 International Eel Symposium,* ed. John M. Casselman and David K. Cairns (Bethesda, MD: American Fisheries Society, 2009), 309–23.

7. Gene S. Helfman and Jennifer B. Clark, "Rotational Feeding: Overcoming Gape-Limited Foraging in Anguillid Eels," *Copeia* 1986 (1986): 679–85.

8. Kenneth Oliveira, "Life History Characteristics and Strategies of the American Eel, *Anguilla rostrata*," *Canadian Journal of Fisheries Aquatic Sciences* 56 (1999): 795–802.

9. Brian M. Jessop, "Geographic Effects on American Eel (*Anguilla rostrata*) Life History Characteristics and Strategies," *Canadian Journal of Fisheries Aquatic Sciences* 67 (2010): 326–46.

10. Oliveira, "Life History," 795–802.

11. Kevin R. Goodwin and Paul L. Angermeier, "Demographic Characteristics of American Eel in the Potomac River Drainage, Virginia," *Transactions of the American Fisheries Society* 132 (2003): 524–35; Derek Wiley, Raymond P. Morgan II, and Robert H. Hilderbrand, "Relations between Physical Habitat and American Eel Abundance in Five River Basins in Maryland," *Transactions of the American Fisheries Society* 133 (2004): 515–26.

12. Stuart A. Welsh and Joni L. Aldinger, "A Semi-automated Method for Monitoring Dam Passage of Upstream Migrant Yellow-Phase American Eels," *North American Journal of Fisheries Management* 34 (2014): 702–9.

13. Joni L. Aldinger and Stuart A. Welsh, "Diel Periodicity and Chronology of Upstream Migration in Yellow-Phase American Eels (*Anguilla rostrata*)," *Environmental Biology of Fishes* 100 (2017): 829–38.

14. Stuart Welsh and Heather Liller, "Environmental Correlates of Upstream Migration of Yellow-Phase American Eels in the Potomac River Drainage," *Transactions of the American Fisheries Society* 142 (2013): 483–91; Stuart A. Welsh, Joni L. Aldinger, Melissa A. Braham, and Jennifer L. Zimmerman, "Synergistic and Singular Effects of River Discharge and Lunar Illumination on Dam Passage of Upstream Migrant Yellow-Phase American Eels," *ICES Journal of Marine Science* 73 (2016): 33–42.

15. Sheila M. Eyler, "Timing and Survival of American Eels Migrating Past Hydroelectric Dams on the Shenandoah River" (PhD diss., West Virginia University, 2014).

16. Sheila M. Eyler, Stuart A. Welsh, David R. Smith, and Mary M. Rockey, "Downstream Passage and Impact of Turbine Shutdowns on Survival of Silver American

Eels at Five Hydroelectric Dams on the Shenandoah River," *Transactions of the American Fisheries Society* 145 (2016): 964–76.

17. Charles A. Wenner and John A. Musick, "Fecundity and Gonad Observation of the American Eel, *Anguilla rostrata*, Migrating from Chesapeake Bay, Virginia," *Journal of the Fisheries Research Board of Canada* 31 (1974): 1387–91; Douglas E. Facey and Michael J. Van Den Avyle, *Species Profiles: Life Histories and Environmental Requirements of Coastal Fishes and Invertebrates (North Atlantic)—American Eel* (Washington, DC: US Fish and Wildlife Service, 1987), 1–28.

CHAPTER 4: SMOKE-SCREEN FORAGING

1. Nikolaas Tinbergen, "Ethology and Stress Diseases," *Science,* n.s., 185 (1974): 20–27.
2. Tinbergen, 20.
3. Nikolaas Tinbergen, "The Curious Behavior of the Stickleback," *Scientific American* 187 (1952): 22–27.
4. "U.S. Pet Ownership Statistics," American Veterinary Medical Association, accessed July 10, 2021, https://www.avma.org/resources-tools/reports-statistics/us-pet-ownership-statistics.
5. Aline H. Kidd and Robert M. Kidd, "Benefits, Problems, and Characteristics of Home Aquarium Owners," *Psychological Reports* 84 (1999): 998–1004.
6. R. Hoogland, Desmond Morris, and Nikolaas Tinbergen, "The Spines of Sticklebacks (*Gasterosteus* and *Pygosteus*) as Means of Defence against Predators (*Perca* and *Esox*)," *Behaviour* 10 (1956): 205–36.
7. Hoogland, Morris, and Tinbergen.
8. Julie E. Schriefer and Melina E. Hale, "Strikes and Startles of Northern Pike (*Esox lucius*): A Comparison of Muscle Activity and Kinematics between S-Start Behaviors," *Journal of Experimental Biology* 207 (2004): 535–44.
9. H. Russ Frith and Robert W. Blake, "The Mechanical Power Output and Hydromechanical Efficiency of Northern Pike (*Esox lucius*) Fast-Starts," *Journal of Experimental Biology* 198 (1995): 1863–73.
10. Hoogland, Morris, and Tinbergen, "Spines of Sticklebacks."
11. David G. Harper and Robert W. Blake, "Prey Capture and Fast-Start Performance of Northern Pike *Esox Lucius*," *Journal of Experimental Biology* 155 (1991): 175–92.
12. Ted Hughes, "Pike," in *Lupercal* (London: Faber and Faber, 1960).
13. P. K. Mishra, "Role of Smokes in Warfare," *Defence Science Journal* 44 (1994): 173–79.
14. Charles S. Bowdoin, "Waterways of Innovation: The Marine Technological Advancements of America's Prohibition Era" (MS thesis, East Carolina University, 2016), 84.
15. Jennifer S. Lewis and William S. Schroeder, "Mud Plume Feeding, a Unique Foraging Behavior of the Bottlenose Dolphin in the Florida Keys," *Gulf of Mexico Science* 21 (2003): 92–97; Leigh G. Torres and Andrew J. Read, "Where to Catch a Fish? The Influence of Foraging Tactics on the Ecology of Bottlenose Dolphins (*Tursiops truncatus*) in Florida Bay, Florida," *Marine Mammal Science* 25 (2009): 797–815.

16. Marcos R. Rossi-Santos and Leonardo L. Wedekin, "Evidence of Bottom Contact Behavior by Estuarine Dolphins (*Sotalia guianensis*) on the Eastern Coast of Brazil," *Aquatic Mammals* 32 (2006): 140–44.

CHAPTER 5: KARPFEN

1. Eugene K. Balon, "Origin and Domestication of the Wild Carp, *Cyprinus carpio*: From Roman Gourmets to the Swimming Flowers," *Aquaculture* 129 (1995): 3–48.
2. Balon.
3. Izaak Walton, *The Compleat Angler or the Contemplative Man's Recreation* (London: S. Printed by T. Maxey for Rich. Marriot, Dunstans Church-Yard Fleetstreet, 1653), 101.
4. Izaak Walton, *The Compleat Angler or the Contemplative Man's Recreation,* 2nd ed. (London: Printed by T. M. for Rich Marriot, Church-Yard Fleetstreet, 1655), 216.
5. Kat Eschner, "This Obscure Fishing Book Is One of the Most Reprinted English Books Ever," *Smithsonian Magazine,* August 9, 2017, https://www.smithsonianmag .com/smart-news/obscure-fishing-book-one-most-reprinted-english-books -ever-180964320/.
6. C. Hart Merriam, "Baird the Naturalist," *Scientific Monthly* 18 (1924): 588–95.
7. Ruthven Deane, "Unpublished Letters of John James Audubon and Spencer F. Baird," *Auk* 23 (1906): 318–34; Ruthven Deane, "Unpublished Letters of John James Audubon and Spencer F. Baird," *Auk* 24 (1907): 53–70.
8. William H. Dall, *Spencer Fullerton Baird: A Biography, Including Selections from His Correspondence with Audubon, Agassiz, Dana, and Others* (London: J. B. Lippincott, 1915).
9. Merriam, "Baird the Naturalist," 588.
10. G. Brown Goode, *A Review of the Fishery Industries of the United States and the Work of the U.S. Fish Commission* (London: William Clowes and Sons, 1883), 55.
11. Goode, 57.
12. Spencer F. Baird, "Report of the Commissioner," in *United States Commission of Fish and Fisheries, Part II, Report of the Commissioner for 1872 and 1873* (Washington, DC: US Government Printing Office, 1874), lxxvi.
13. A. Woldt, "Die besetzung des Stettiner Haffs mit karpfen," in *Circulare des Deutschen Fischerei-Vereins im Jahre 1881* (Berlin: W. Moeser Hofbuchdruckerei, 1882), 232–37; translated from German.
14. Spencer F. Baird, "Report of the Commissioner," in *United States Commission of Fish and Fisheries, Part III, Report of the Commissioner for 1873–4 and 1874–5* (Washington, DC: US Government Printing Office, 1876), xxxvi.
15. Leon J. Cole, "The German Carp in the United States," in *Appendix to the Report of the Commissioner of Fisheries to the Secretary of Commerce and Labor for the Year Ending June 30, 1904* (Washington, DC: US Government Printing Office, Department of Commerce and Labor, Bureau of Fisheries, 1905), 544–45.
16. Charles W. Smiley, "Answers to 118 Questions Relative to German Carp," *Bulletin of the United States Fish Commission* 3 (1883): 241–48.
17. Smiley.
18. Cole, "German Carp in the United States."

19. Ralph S. Tarr, "The United States Fish Commission," *Nature* 31 (1884): 128-30.

20. Cole, "German Carp in the United States," 546.

21. H. B. Davis, "Gratifying Results of Propagating German Carp—Bream and Carp in Ponds Together—Table Qualities of Carp," *Bulletin of the United States Fish Commission* 2 (1883): 317-18.

22. Edward Thompson, "Edible Qualities of Carp," *Bulletin of the United States Fish Commission* 4 (1884): 176.

23. Smiley, "118 Questions Relative to German Carp," 242-43.

24. Charles W. Smiley, "Notes on the Edible Qualities of German Carp and Hints about Cooking Them," *Bulletin of the United States Fish Commission* 3 (1883): 305-32.

25. Smiley, 305.

26. Smiley, 307.

27. Smiley, 306.

28. Rudolph Hessel, "The Carp, and Its Culture in Rivers and Lakes; and Its Introduction into America," in *United States Commission of Fish and Fisheries, Part IV, Report of the Commissioner for 1875-1876* (Washington, DC: US Government Printing Office, 1878), 865-900.

29. John S. Gottschalk, "The Introduction of Exotic Animals into the United States," in *Towards a New Relationship of Man and Nature in Temperate Lands* (Morges, Switzerland: International Union for Conservation of Nature and Natural Resources, 1967), 124-40.

30. A. Wilson Armistead, "A Transfer of Leather Carp (*Cyprinus carpio*) from the Government Ponds at Washington, U.S.A., to Scotland," *Bulletin of the United States Fish Commission* 1 (1882): 341-42.

31. Goode, *Fishery Industries of the United States,* 73.

32. Cole, "German Carp in the United States," 548.

33. C. W. Threinen and William T. Helm, "Experiments and Observations Designed to Show Carp Destruction of Aquatic Vegetation," *Journal of Wildlife Management* 18 (1954): 247-51; S. A. Miller and T. A. Crowl, "Effects of Common Carp (*Cyprinus carpio*) on Macrophytes and Invertebrate Communities in a Shallow Lake," *Freshwater Biology* 51 (2006): 85-94.

34. Gottschalk, "Introduction of Exotic Animals," 131.

35. Alain J. Crivelli, "The Destruction of Aquatic Vegetation by Carp: A Comparison between Southern France and the United States," *Hydrobiologia* 106 (1983): 37-41.

36. Vanessa L. Lougheed, Barb Crosbie, and Patricia Chow-Fraser, "Predictions on the Effect of Common Carp (*Cyprinus carpio*) Exclusion on Water Quality, Zooplankton, and Submergent Macrophytes in a Great Lakes Wetland," *Canadian Journal of Fisheries and Aquatic Sciences* 55 (1998): 1189-97.

37. Joseph L. Bonneau and Dennis L. Scarnecchia, "The Zooplankton Community of a Turbid Great Plains (USA) Reservoir in Response to a Biomanipulation with Common Carp (*Cyprinus carpio*)," *Transactions of the Kansas Academy of Science* 117 (2014): 181-92.

38. Miller and Crowl, "Effects of Common Carp"; Joseph J. Parkos III, Victor J. Santucci Jr., and David H. Wahl, "Effects of Adult Common Carp (*Cyprinus carpio*) on Multiple Trophic Levels in Shallow Mesocosms," *Canadian Journal of Fisheries*

and Aquatic Sciences 60 (2003): 182–92; Mark A. Kaemingk et al., "Common Carp Disrupt Ecosystem Structure and Function through Middle-Out Effects," *Marine and Freshwater Research* (2016): https://doi.org/10.1071/MF15068.

39. Charles W. Smiley, "Brief Notes upon Fish and the Fisheries," *Bulletin of the United States Fish Commission* 4 (1884): 305–20.

40. William E. Meehan, "To the Committee of Fish and Game of the Blooming Grove Park Association," in *Report of the Commissioners of Fisheries of the State of Pennsylvania for the Year 1900* (Harrisburg: Wm. Stanley Ray, State Printer of Pennsylvania, 1900), 138.

41. Richard Cotchefer, "Report of the General Foreman of Hatcheries," in *Seventh Report of the Forest, Fish, and Game Commission of the State of New York* (Albany, NY: J. B. Lyon, 1902), 61.

42. J. M. Willson Jr., "Florida Fur-Farming," *Bulletin of the United States Fish Commission* 17 (1898): 369–71.

43. David Starr Jordan, "Spencer Fullerton Baird and the United States Fish Commission," *Scientific Monthly* 17 (1923): 97–107.

44. Jordan, 104.

45. Cole, "German Carp in the United States," 544.

46. Cole, 636.

CHAPTER 6: HORNYHEAD

1. Robert E. Jenkins and Noel M. Burkhead, *Freshwater Fishes of Virginia* (Bethesda, MD: American Fisheries Society, 1994), 303.

2. Robert E. Lennon and Phillip S. Parker, "The Stoneroller, *Campostoma anomalum* (Rafinesque), in Great Smoky Mountains National Park," *Transactions of the American Fisheries Society* 89 (1960): 263–70.

3. Tyson R. Roberts, "Unculi (Horny Projections Arising from Single Cells), an Adaptive Feature of the Epidermis of Ostariophysan Fishes," *Zoologica Scripta* 11 (1982): 55–76.

4. Jacob Reighard, "The Function of the Pearl Organs of Cyprinidae," *Science* 17 (1903): 531; Jacob Reighard, "The Pearl Organs of American Minnows in Their Relation to the Factors of Descent," *Science,* n.s., 31 (1910): 472; Norman H. Stewart, "Pearl Organs of the Cyprinidae," *Proceedings of the Pennsylvania Academy of Science* 27 (1953): 221–24; Ernest A. Lachner and Robert E. Jenkins, "Systematics, Distribution, and Evolution of the Chub Genus *Nocomis* (Cyprinidae) in the Southwestern Ohio River Basin, with the Description of a New Species," *Copeia* 1967 (1967): 557–80.

5. Fr. Maurer, *Die epidermis und ihre abkömmlinge* (Leipzig: Wilhelm Engelmann, 1895), 92.

6. Martin L. Wiley and Bruce B. Collette, "Breeding Tubercles and Contact Organs in Fishes: Their Occurrence, Structure, and Significance," *Bulletin of the American Museum of Natural History* 143 (1970): 202.

7. Nancy J. Alexander, "Comparison of a and b Keratin in Reptiles," *Zeitschrift für Zellforschung und mikroskopische Anatomie* 110 (1970): 153–65.

8. Richard H. C. Bonser and Peter P. Purslow, "The Young's Modulus of Feather Keratin," *Journal of Experimental Biology* 198 (1995): 1029–33.

9. Leopold Eckhart et al., "Identification of Reptilian Genes Encoding Hair Keratin–Like Proteins Suggests a New Scenario for the Evolutionary Origin of Hair," *Proceedings of the National Academy of Sciences* 105 (2008): 18419–23.

10. William R. McGuire, "Relationships of Epidermal Morphology and Breeding Behaviors in Pebble Nest-Building Minnows (Pisces: Cyprinidae)" (MS thesis, University of Richmond, 1993).

11. Ernest A. Lachner, "Studies of the Biology of the Cyprinid Fishes of the Chub Genus *Nocomis* of Northeastern United States," *American Midland Naturalist* 48 (1952): 433–66; Ernest A. Lachner and Robert E. Jenkins, "Systematics, Distribution, and Evolution of the Chub Genus *Nocomis* (Cyprinidae) in the Southwestern Ohio River Basin, with the Description of a New Species," *Copeia* 1967 (1967): 557–80.

12. Jacob Reighard, "Methods of Studying the Habitats of Fishes, with an Account of the Breeding Habitats of the Horned Dace," *Bulletin of the U.S. Bureau of Fisheries* 28 (1910): 1111–36.

13. Rudolph J. Miller, "Reproductive Behavior of the Stoneroller Minnow, *Campostoma anomalum pullum*," *Copeia* (1962): 407–17.

14. Mark H. Sabaj, "Spawning Clasps and Gamete Deposition in Pebble Nest-Building Minnows (Pisces: Cyprinidae)" (MS thesis, University of Richmond, 1992), 50.

15. Reighard, "Breeding Habitats of the Horned Dace," 1128–29.

16. Miller, "Reproductive Behavior of the Stoneroller," 413.

17. Michael R. Ross, "The Breeding Behavior and Hybrid Potential of the Northern Creek Chub *Semotilus atromaculatus atromaculatus* (Mitchill)" (PhD diss., Ohio State University, 1975), 30.

18. Ross, 60–61.

19. Reighard, "Pearl Organs of Cyprinidae," 531.

20. Reighard, "Breeding Habitats of the Horned Dace," 1131.

21. Sabaj, "Spawning Clasps"; Mark H. Sabaj, Eugene G. Maurakis, and William S. Woolcott, "Spawning Behaviors in the Bluehead Chub, *Nocomis leptocephalus*, River Chub, *N. micropogon* and Central Stoneroller, *Campostoma anomalum*," *American Midland Naturalist* 144 (2000): 187–201.

22. Lachner, "Nocomis of Northeastern United States"; Lachner and Jenkins, "Chub Genus *Nocomis*."

23. McGuire, "Pebble Nest-Building Minnows," 17; Miller, "Reproductive Behavior of the Stoneroller."

24. McGuire, "Pebble Nest-Building Minnows," 17.

CHAPTER 7: NOCOMIS NESTS

1. Jacob Reighard, "The Breeding Habits of the River Chub, *Nocomis micropogon* (Cope)," *Papers of the Michigan Academy of Science, Arts and Letters* 28 (1943): 393–423.

2. David J. Schmidly, "What It Means to Be a Naturalist and the Future of Natural History at American Universities," *Journal of Mammalogy* 86 (2005): 449–56.

3. Jonathan W. Moore, "Animal Ecosystem Engineers in Streams," *Bioscience* 56 (2006): 237–46.

4. Justin P. Wright, Clive G. Jones, and Alexander S. Flecker, "An Ecosystem Engineer, the Beaver, Increases Species Richness at the Landscape Scale," *Oecologia* 132 (2002): 96–101.

5. L. Scott Mills, Michael E. Soulé, and Daniel F. Doak, "The Keystone-Species Concept in Ecology and Conservation," *BioScience* 43 (1993): 219–24.

6. Clive G. Jones, John H. Lawton, and Moshe Shachak, "Positive and Negative Effects of Organisms as Physical Ecosystem Engineers," *Ecology* 78 (1997): 1946–57.

7. Jacob Reighard, "Methods of Studying the Habitats of Fishes, with an Account of the Breeding Habitats of the Horned Dace," *Bulletin of the U.S. Bureau of Fisheries* 28 (1910): 1111–36.

8. Reighard, "Breeding Habitats of the River Chub," 399.

9. Reighard, 402.

10. Reighard, 403.

11. Reighard, 403.

12. Reighard, 405.

13. Reighard, 409.

14. Reighard, 405–6.

15. Reighard, 413.

16. Reighard, 419.

17. Carol E. Johnston, "The Benefit to Some Minnows of Spawning in the Nests of Other Species," *Environmental Biology of Fishes* 40 (1994): 213–18; Carol E. Johnston, "Nest Association in Fishes: Evidence for Mutualism," *Behavioral Ecology and Sociobiology* 35 (1994): 379–83.

18. Ernest A. Lachner, "Studies of the Biology of the Cyprinid Fishes of the Chub Genus *Nocomis* of Northeastern United States," *American Midland Naturalist* 48 (1952): 433–66.

19. George F. Turner and Tony J. Pitcher, "Attack Abatement: A Model for Group Protection by Combined Avoidance and Dilution," *American Naturalist* 128 (1986): 228–40.

20. William D. Hamilton, "Geometry for the Selfish Herd," *Journal of Theoretical Biology* 31 (1971): 295–311.

21. Carol E. Johnston, "Spawning Activities of *Notropis chlorocephalus, Notropis chiliticus,* and *Hybopsis hypsinotus,* Nest Associates of *Nocomis leptocephalus* in the Southeastern United States, with Comments on Nest Association (Cypriniformes, Cyprinidae)," *Brimleyana* 17 (1991): 77–88.

22. Jeff Ollerton, "'Biological Barter': Patterns of Specialization Compared across Different Mutualisms," in *Plant-Pollinator Interactions: From Specialization to Generalization* (Chicago: University of Chicago Press, 2006), 411–35.

23. Rudolph J. Miller, "Behavior and Ecology of Some North American Cyprinid Fishes," *American Midland Naturalist* 72 (1964): 313–57; Eugene G. Maurakis, "Breeding Behaviors in *Nocomis platyrhynchus* and *Nocomis raneyi* (Actinopterygii: Cyprinidae)," *Virginia Journal of Science* 49 (1998): 227–36; Eugene G. Maurakis, William S. Woolcott, and Mark H. Sabaj, "Reproductive-Behavioral Phylogenetics of Nocomis Species-Groups," *American Midland Naturalist* 126 (1991): 103–10; Mark H. Sabaj, Eugene G. Maurakis, and William S. Woolcott, "Spawning Behaviors in the Bluehead Chub, *Nocomis leptocephalus,* River Chub, *Nocomis micropogon* and Central Stoneroller, *Campostoma anomalum,*" *American Midland Naturalist* 144 (2000): 187–201; Eugene G. Maurakis and Terre D. Green, "Comparison of

Spawning and Non-spawning Substrates in Nests of Species of *Exoglossum* and *Nocomis* (Actinopterygii: Cyprinidae)," *Virginia Journal of Science* 52 (2001): 25–34; Brandon K. Peoples, Ryan A. McManamay, Donald J. Orth, and Emmanuel A. Frimpong, "Nesting Habitat Use by River Chubs in a Hydrologically Variable Appalachian Tailwater," *Ecology of Freshwater Fish* 23 (2014): 283–93.

24. Ernest A. Lachner and Robert E. Jenkins, "Systematics, Distribution, and Evolution of the Chub Genus *Nocomis* Girard (Pisces, Cyprinidae) of the Eastern United States, with Descriptions of New Species," *Smithsonian Contributions to Zoology* 85 (1971): 1–97; Ernest A. Lachner and Robert E. Jenkins, "Systematics, Distribution, and Evolution of the *Nocomis biguttatus* Species Group (Family Cyprinidae: Pisces) with a Description of a Species from the Ozark Upland," *Smithsonian Contributions to Zoology* 91 (1971): 1–28.

25. Michael R. Ross and Roger J. Reed, "The Reproductive Behavior of the Fallfish *Semotilus corporalis*," *Copeia* 1978 (1978): 215–21.

26. Reighard, "Breeding Habits of the Horned Dace."

27. Maurakis and Green, "Species of *Exoglossum* and *Nocomis*," 25–34; Evelyn M. van Duzer, "Observations on the Breeding Habits of the Cut-Lips Minnow, *Exoglossum maxillingua*," *Copeia* 1939 (1939): 65–75.

CHAPTER 8: SPOTS AND DOTS

1. John W. Francis, *Reminiscences of Samuel Latham Mitchill, M.D., LL.D* (New York: John F. Trow, 1859), 1–31.

2. Joel K. Mead, *The National Register,* vol. 4, no. 8 (Washington City: Printed and Published by the Proprietor, 1817), 121.

3. William D. Johnston, *History of the Library of Congress,* vol. 1, *1800–1864* (Washington, DC: US Government Printing Office, 1904), 45.

4. Johnston, 45.

5. Thomas C. Danisi, "Preserving the Legacy of Meriwether Lewis: The Letters of Samuel Latham Mitchill," *Quarterly Journal of the Lewis and Clark Foundation* 36 (2010): 8–11.

6. Francis, *Samuel Latham Mitchill,* 22; Danisi, "Legacy of Meriwether Lewis."

7. Samuel L. Mitchill, *Report, in Part, on the Fishes of New York, 1814,* ed. Theodore Gill (Washington, DC: Printed for the editor, 1898).

8. Samuel L. Mitchill, "Memoir on Ichthyology: The Fishes of New York, Described and Arranged," *American Monthly Magazine and Critical Review* 2 (1818): 321–28.

9. Tony J. Pitcher and Julia K. Parrish, "Functions of Shoaling Behaviour in Teleosts," in *The Behaviour of Teleost Fishes,* ed. Tony J. Pitcher, 2nd ed. (London: Croom Helm, 1993), 364–439.

10. Johann Delcourt and Pascal Poncin, "Shoals and Schools: Back to the Heuristic Definitions and Quantitative References," *Reviews in Fish Biology and Fisheries* 22 (2012): 595–619.

11. Noam Y. Miller and Robert Gerlai, "Shoaling in Zebrafish: What We Don't Know," *Reviews in the Neurosciences* 22 (2011): 17–25.

12. Jens Krause, Jean-Guy J. Godin, and Dan Rubenstein, "Group Choice as a Function of Group Size Differences and Assessment Time in Fish: The Influence of Species Vulnerability to Predation," *Ethology* 104 (1998): 68–74.

13. William D. Hamilton, "Geometry for the Selfish Herd," *Journal of Theoretical Biology* 31 (1971): 295–311; Lesley J. Morrell, Graeme D. Ruxton, and Richard James, "Spatial Positioning in the Selfish Herd," *Behavioral Ecology* 22 (2010): 16–22.

14. Jacob Reighard, "Methods of Studying the Habitats of Fishes, with an Account of the Breeding Habitats of the Horned Dace," *Bulletin of the U.S. Bureau of Fisheries* 28 (1910): 1111–36.

15. Mitchill, "Memoir on Ichthyology," 325.

16. Dr. Seuss, *Oh Say Can You Say?* (New York: Random House, 1979).

17. Anonymous, "Scientific News," *English Mechanic and World of Science* 21 (1875): 10.

18. Anonymous, "Cause of the Black Spots on the Scales of Fish," in *Annual Record of Science and Industry for 1876, Edited by Spencer F. Baird with the Assistance of Eminent Men of Science* (New York: Harper and Brothers, 1877), 332.

19. Glenn L. Hoffman, *Synopsis of Strigeoidea (Trematoda) of Fishes and Their Life Cycles, Fishery Bulletin 175* (Washington, DC: US Government Printing Office, 1960), 439–69; Glenn L. Hoffman, *Parasites of North American Freshwater Fishes,* 2nd ed. (Ithaca, NY: Cornell University Press, 1999), 191–97.

20. Tim M. Berra and Ray-Jean Au, "Incidence of Black Spot Disease in Fishes in Cedar Fork Creek, Ohio," *Ohio Journal of Science* 78 (1978): 318–22.

21. Michael Tobler and Ingo Schlupp, "Influence of Black Spot Disease on Shoaling Behaviour in Female Western Mosquitofish, *Gambusia affinis* (Poeciliidae, Teleostei)," *Environmental Biology of Fishes* 81 (2008): 29–34.

22. Jens Krause and Jean-Guy J. Godin, "Influence of Parasitism on Shoal Choice in the Banded Killifish (*Fundulus diaphanous,* Teleostei, Cyprinodontidae)," *Ethology* 102 (1996): 40–49.

23. D. Bumann, J. Krause, and D. Rubenstein, "Mortality Risk of Spatial Positions in Animal Groups: The Danger of Being in the Front," *Behaviour* 134 (1997): 1063–76.

24. William D. Hamilton and Marlene Zuk, "Heritable True Fitness and Bright Birds: A Role for Parasites?," *Science* 218 (1982): 384–87.

25. Manfred Milinski and Theo C. M. Bakker, "Female Sticklebacks Use Male Coloration in Mate Choice and Hence Avoid Parasitized Males," *Nature* 344 (1990): 330–33; Iain Barber, Stephen A. Arnott, Victoria A. Braithwaite, Jennifer Andrew, and Felicity A. Huntingford, "Indirect Fitness Consequences of Mate Choice in Sticklebacks: Offspring of Brighter Males Grow Slowly but Resist Parasitic Infections," *Proceedings of the Royal Society of London B: Biological Sciences* 268 (2001): 71–76.

CHAPTER 9: ANOMALUS ANATOMY

1. Edward D. Cope, "Synopsis of the Cyprinidae of Pennsylvania," *Transactions of the American Philosophical Society,* n.s., 13, part 3, art. 13 (1869): 353.

2. Cope, "Cyprinidae of Pennsylvania," 394.

3. Walter C. Kraatz, "The Intestine of the Minnow *Campostoma anomalum* (Rafinesque), with Special Reference to the Development of Its Coiling," *Ohio Journal of Science* 24 (1924): 265–98; Mary D. Rogick, "Studies on the Comparative Histology of the Digestive Tube of Certain Teleost Fishes, II. A Minnow (*Campostoma anomalum*)," *Journal of Morphology and Physiology* 52 (1931): 1–25.

4. William J. Matthews, Mary E. Power and Arthur J. Stewart, "Depth Distribution of Campostoma Grazing Scars in an Ozark Stream," *Environmental Biology of Fishes* 17 (1986): 292.

5. Stuart G. Fisher and Gene E. Likens, "Energy Flow in Bear Brook, New Hampshire: An Integrative Approach to Stream Ecosystem Metabolism," *Ecological Monographs* 43 (1973): 421–39.

6. Cope, "Cyprinidae of Pennsylvania," 396.

7. Edward D. Cope, "On the Distribution of Fresh-Water Fishes in the Allegheny Region of Southwestern Virginia," *Journal of the Academy of Natural Sciences of Philadelphia,* ser. 2, 6, part 3, art. 5 (1869): 210.

8. James H. Johnson and David S. Dropkin, "Piscivory by the Central Stoneroller *Campostoma anomalum,*" *Journal of the Pennsylvania Academy of Science* 66, no. 2 (1992): 90–91.

9. James F. Fowler and Charles A. Taber, "Food Habits and Feeding Periodicity in Two Sympatric Stonerollers (Cyprinidae)," *American Midland Naturalist* 113 (1985): 217–24.

10. A. K. Ray, K. Ghosh, and E. Ringø, "Enzyme-Producing Bacteria Isolated from Fish Gut: A Review," *Aquaculture Nutrition* 18 (2012): 465–92.

11. Hannah Grice, "Effects on *Campostoma oligolepis* Digestive Morphology and Gut Microbiota Composition across a Gradient of Urbanization" (MS thesis, Kennesaw State University, 2017), 31.

12. Donovan P. German, "Do Herbivorous Minnows Have 'Plug-Flow Reactor' Guts? Evidence from Digestive Enzyme Activities, Gastrointestinal Fermentation, and Luminal Nutrient Concentrations," *Journal of Comparative Physiology B* 179 (2009): 759–71.

13. D. W. Rimmer and W. J. Wiebe, "Fermentative Microbial Digestion in Herbivorous Fishes," *Journal of Fish Biology* 31 (1987): 229–36.

14. Mary E. Power and William J. Matthews, "Algae-Grazing Minnows (*Campostoma anomalum*), Piscivorous Bass (*Micropterus* spp.), and the Distribution of Attached Algae in a Small Prairie-Margin Stream," *Oecologia* 60 (1983): 328–32.

15. Chad W. Hargrave et al., "Indirect Food Web Interactions Increase Growth of an Algivorous Stream Fish," *Freshwater Biology* 51 (2006): 1901–10.

16. William P. Coffman, Kenneth W. Cummins, and John C. Wuycheck, "Energy Flow in a Woodland Stream Ecosystem: I. Tissue Support Trophic Structure of the Autumnal Community," *Archiv für Hydrobiologie* 68 (1971): 232–76.

17. Coffman, Cummins, and Wuycheck.

18. Cope, "Fishes in the Allegheny Region," 209.

CHAPTER 10: SUCKER SAVVY

1. *Merriam-Webster Dictionary,* s.v. "bottom-feeder," accessed June 10, 2019, https://www.merriam-webster.com/dictionary/bottom-feeder.

2. Mark Holey, Bruce Hollender, Mark Imhof, Roman Jesien, Richard Konopacky, Michael Toneys, and Daniel Coble, "Never Give a Sucker an Even Break," *Fisheries* 4, no. 1 (1979): 2–6; Steven J. Cooke, Christopher M. Bunt, Steven J. Hamilton, Cecil A. Jennings, Michael P. Pearson, Michale S. Cooperman, and Douglas F. Markle, "Threats, Conservation Strategies, and Prognosis for Suckers

(Catostomidae) in North America: Insights from Regional Case Studies of a Diverse Family of Non-game Fishes," *Biological Conservation* 121 (2005): 317–31.

3. Lawrence M. Page and Carol E. Johnston, "Spawning in the Creek Chubsucker, *Erimyzon oblongus,* with a Review of Spawning Behavior in Suckers (Catostomidae)," *Environmental Biology of Fishes* 27 (1990): 265.

4. Page and Johnston, "Spawning Behavior in Suckers," 265–66.

5. Scott M. Gende, Richard T. Edwards, Mary F. Willson, and Mark S. Wipfli, "Pacific Salmon in Aquatic and Terrestrial Ecosystems: Pacific Salmon Subsidize Freshwater and Terrestrial Ecosystems through Several Pathways, Which Generates Unique Management and Conservation Issues but Also Provides Valuable Research Opportunities," *BioScience* 52 (2002): 917–28.

6. Robert J. Naiman, Robert E. Bilby, Daniel E. Schindler, and James M. Helfield, "Pacific Salmon, Nutrients, and the Dynamics of Freshwater and Riparian Ecosystems," *Ecosystems* 5 (2002): 399–417.

7. Scott D. Tiegs, Peter S. Levi, Janine Ruegg, Dominic T. Chaloner, Jennifer L. Tank, and Gary A. Lamberti, "Ecological Effects of Live Salmon Exceed Those of Carcasses during an Annual Spawning Migration," *Ecosystems* 14 (2011): 598–614.

8. Evan S. Childress, J. David Allan, and Peter B. McIntyre, "Nutrient Subsidies from Iteroparous Fish Migrations Can Enhance Stream Productivity," *Ecosystems* 17 (2014): 522–34.

9. Robert E. Jenkins and Noel M. Burkhead, *Freshwater Fishes of Virginia* (Bethesda, MD: American Fisheries Society, 1994), 461.

10. Linden F. Edwards, "The Protactile Apparatus of the Mouth of the Catostomid Fishes," *Anatomical Record* 33 (1926): 257.

11. Rudolph J. Miller and Howard E. Evans, "External Morphology of the Brain and Lips in Catostomid Fishes," *Copeia* 1965 (1965): 467–87.

12. Edwards, "Mouth of the Catostomid Fishes," 257.

13. *Merriam-Webster Dictionary,* s.v. "sucker," accessed June 10, 2019, https://www .merriam-webster.com/dictionary/sucker.

14. John R. Greeley, "The Spawning Habits of Brook, Brown, and Rainbow Trout and the Problem of Egg Predation," *Transactions of the American Fisheries Society* 62 (1932): 239.

15. Greeley, 247–48.

16. Greeley, 245.

17. Greeley, 248.

18. Fritz H. Johnson, "Responses of Walleye (*Stizostedion vitreum vitreum*) and Yellow Perch (*Perca flavescens*) Populations to Removal of White Sucker (*Catostomus commersoni*) from a Minnesota Lake, 1966," *Journal of the Fisheries Board of Canada* 34 (1977): 1633–42; Daniel B. Hayes, William W. Taylor, and James C. Schneider, "Response of Yellow Perch and the Benthic Invertebrate Community to a Reduction in the Abundance of White Suckers," *Transactions of the American Fisheries Society* 121 (1992): 36–53.

19. Jenkins and Burkhead, *Freshwater Fishes of Virginia,* 460.

20. Milton L. Bowman, "Life History of the Black Redhorse, *Moxostoma duquesnei* (Lesueur), in Missouri," *Transactions of the American Fisheries Society* 99 (1970): 546–59.

21. J. A. Baker and S. A. Foster, "Observations on a Foraging Association between Two Freshwater Stream Fishes," *Ecology of Freshwater Fish* 3 (1994): 137–79; Fabricio

B. Teresa and Fernando R. Carvalho, "Feeding Association between Benthic and Nektonic Neotropical Stream Fishes," *Neotropical Ichthyology* 6 (2008): 109–11; Fabricio B. Teresa, Cristina Sazima, Ivan Sazima, and Sergio R. Floeter, "Predictive Factors of Species Composition of Follower Fishes in Nuclear-Follower Feeding Associations: A Snapshot Study," *Neotropical Ichthyology* 12 (2014): 913–19.

22. José Sabino, Luciana P. Andrade, Ivan Sazima, Fabricio B. Teresa, Sergio R. Floeter, Cristina Sazima, and Roberta M. Bonaldo, "Following Fish Feeding Associations in Marine and Freshwater Habitats," *Marine and Freshwater Research* 68 (2017): 381–87.

23. S. Fiedel, "Older Than We Thought: Implications of Corrected Dates for Paleoindians," *American Antiquity* 64 (1999): 95–115.

24. Renee B. Walker and Boyce N. Driskell, *Foragers of the Terminal Pleistocene in North America* (Lincoln: University of Nebraska Press, 2007); Renee B. Walker, Kandace R. Detwiler, Scott C. Meeks, and Boyce N. Driskell, "Berries, Bones, and Blades: Reconstructing Late Paleoindian Subsistence Economy at Dust Cave, Alabama," *Midcontinental Journal of Archaeology* 26 (2001): 169–97.

25. Mark Q. Sutton, "The 'Fishing Link': Salmonids and the Initial Peopling of the Americas," *PaleoAmerica* 3 (2017): 231–59.

26. D. M. Morris and L. E. Dawson, "Storage Stability of Mechanically Deboned Sucker (Catostomidae) Flesh," *Journal of Food Science* 44 (1979): 1093–96; Mark C. Nilsson, "Market Alternatives for Carp, Alewife, and Sucker from the Michigan Great Lakes" (report prepared for Michigan State University, 1979), accessed June 6, 2019, https://ageconsearch.umn.edu/record/11112; J. Mai and J. E. Kinsella, "Composition of Lipids and Proteins of Deboned Minced and Filleted White Sucker (*Catostomus commersoni*)," *Journal of Food Biochemistry* 3 (1980): 229–39.

27. Jenkins and Burkhead, *Freshwater Fishes of Virginia*, 462.

CHAPTER 11: MADTOM MINIATURES

1. William R. Taylor, *A Revision of the Catfish Genus Noturus Rafinesque, with an Analysis of Higher Groups in the Ictaluridae,* Bulletin of the United States National Museum, no. 282 (Washington, DC: Smithsonian Institution Press, 1969), plates 3 and 4.

2. Carl L. Hubbs and Claude W. Hibbard, "*Ictalurus lambda,* a New Catfish, Based on a Pectoral Spine from the Lower Pliocene of Kansas," *Copeia* 1951 (1951): 8–14.

3. Michael L. Fine et al., "Pectoral Spine Locking and Sound Production in the Channel Catfish *Ictalurus punctatus,*" *Copeia* (1997): 777–90.

4. Hugh D. Reed, "The Poison Glands of *Noturus* and *Schilbeodes,*" *American Naturalist* 41 (1907): 553–66.

5. William S. Birkhead, "Toxicity of Stings of Ariid and Ictalurid Catfishes," *Copeia* 1972 (1972): 790–807.

6. L. Scott Forbes, "Prey Defences and Predator Handling Behaviour: The Dangerous Prey Hypothesis," *Oikos* 55 (1989): 155–58.

7. Forbes, 155–58.

8. Philip A. Cochran and Mark A. Zoller, "'Willow Cats' for Sale? Madtoms (Genus *Noturus*) as Bait in the Upper Mississippi River Valley," *American Currents* 35,

no. 2 (2009): 1–8; Jeremy J. Wright, "Diversity, Phylogenetic Distribution, and Origins of Venomous Catfishes," *BMC Evolutionary Biology* 9 (2009): 282.

9. Wright, "Origins of Venomous Catfishes," 282.

10. Bishop T. Bosher, Scott H. Newton, and Michael L. Fine, "The Spines of the Channel Catfish, *Ictalurus punctatus,* as an Anti-predator Adaptation: An Experimental Study," *Ethology* 112 (2005): 188–95.

11. Bridget Emmett and Philip A. Cochran, "The Response of a Piscivore (*Micropterus salmoides*) to a Venomous Prey Species (*Noturus gyrinus*)," *Journal of Freshwater Ecology* 25 (2010): 475–79.

12. David H. Baker, "An Unusual Foreign Body: Catfish Spine," *Pediatric Radiology* 27 (1997): 585.

13. Bruce W. Halstead, Leonard S. Kuninobu, and Harold G. Hebard, "Catfish Stings and the Venom Apparatus of the Mexican Catfish, *Galeichthys felis* (Linnaeus)," *Transactions of the American Microscopical Society* 72 (1953): 297–314.

14. Jarrod E. Church and Wayne C. Hodgson, "The Pharmacological Activity of Fish Venoms," *Toxicon* 40 (2002): 1083–93.

15. Donald M. McKinstry, "Catfish Stings in the United States: Case Report and Review," *Wilderness and Environmental Medicine* 4 (1993): 293–303.

16. Jacob J. D. Egge and Andrew M. Simons, "Evolution of Venom Delivery Structures in Madtom Catfishes (Siluriformes: Ictaluridae)," *Biological Journal of the Linnean Society* 102 (2011): 115–29; Jeremy J. Wright, "Evolutionary History of Venom Glands in the Siluriformes," in *Evolution of Venomous Animals and Their Toxins* (Netherlands: Springer, 2015), 1–19.

17. Wright, "Origins of Venomous Catfishes," 282.

18. William L. Smith and Ward C. Wheeler, "Venom Evolution Widespread in Fishes: A Phylogenetic Road Map for the Bioprospecting of Piscine Venoms," *Journal of Heredity* 97 (2006): 206–17.

19. Gisha Sivan, "Fish Venom: Pharmacological Features and Biological Significance," *Fish and Fisheries* 10 (2009): 159–72; Church and Hodgson, "Activity of Fish Venoms."

CHAPTER 12: TREES AND TROUT

1. Franklin Toker, *Fallingwater Rising: Frank Lloyd Wright, E. J. Kaufmann, and America's Most Extraordinary House* (New York: Alfred A. Knopf, 2003); Robert W. Weisberg, "Frank Lloyd Wright's Fallingwater: A Case Study in Inside-the-Box Creativity," *Creativity Research Journal* 23 (2011): 296–312.

2. Daniel M. Evans, W. Michael Aust, C. Andrew Dolloff, Ben S. Templeton, and John A. Peterson, "Eastern Hemlock Decline in Riparian Areas from Maine to Alabama," *Northern Journal of Applied Forestry* 28 (2011): 97–104; Kurt J. Krapfl, Eric J. Holzmueller, and Michael A. Jenkins, "Early Impacts of Hemlock Woolly Adelgid in *Tsuga canadensis* Forest Communities of the Southern Appalachian Mountains," *Journal of the Torrey Botanical Society* 138 (2011): 93–106; Evan L. Preisser, Kelly L. F. Oten, and Fred P. Hain, "Hemlock Woolly Adelgid in the Eastern United States: What Have We Learned?," *Southeastern Naturalist* 13 (2014): 1–15.

3. Scott R. Abella, "Impacts and Management of Hemlock Woolly Adelgid in National Parks of the Eastern United States," *Southeastern Naturalist* 13 (2014): 16–45.

4. Aaron M. Ellison et al., "Loss of Foundation Species: Consequences for the Structure and Dynamics of Forested Ecosystems," *Frontiers in Ecology and the Environment* 3 (2005): 479–86; Katherine L. Martin and P. Charles Goebel, "The Foundation Species Influence of Eastern Hemlock (*Tsuga canadensis*) on Biodiversity and Ecosystem Function on the Unglaciated Allegheny Plateau," *Forest Ecology and Management* 289 (2013): 143–52.

5. Julian L. Hadley, "Understory Microclimate and Photosynthetic Response of Saplings in an Old-Growth Eastern Hemlock (*Tsuga canadensis* L.) Forest," *Ecoscience* 7 (2000): 66–72.

6. S. Catovsky, N. M. Holbrook, and F. A. Bazzaz, "Coupling Whole-Tree Transpiration and Canopy Photosynthesis in Coniferous and Broad-Leaved Tree Species," *Canadian Journal of Forest Research* 32 (2002): 295–309; Michael J. Daley, Nathan G. Phillips, Cory Pettijohn, and Julian L. Hadley, "Water Use by Eastern Hemlock (*Tsuga canadensis*) and Black Birch (*Betula lenta*): Implications of Effects of the Hemlock Woolly Adelgid," *Canadian Journal of Forest Research* 37 (2007): 2031–40; Steven T. Brantley, Chelcy Ford Miniat, Katherine J. Elliott, Stephanie H. Laseter, and James M. Vose, "Changes to Southern Appalachian Water Yield and Stormflow after Loss of a Foundation Species," *Ecohydrology* 8 (2015): 518–28.

7. J. Todd Petty, Jeff L. Hansbarger, Brock M. Huntsman, and Patricia M. Mazik, "Brook Trout Movement in Response to Temperature, Flow, and Thermal Refugia within a Complex Appalachian Riverscape," *Transactions of the American Fisheries Society* 141 (2012): 1060–73.

8. Robert E. Jenkins and Noel M. Burkhead, *Freshwater Fishes of Virginia* (Bethesda, MD: American Fisheries Society, 1994), 577.

9. John A. Sweka and Kyle J. Hartman, "Effects of Large Woody Debris Addition on Stream Habitat and Brook Trout Populations in Appalachian Streams," *Hydrobiologia* 559 (2006): 363–78; Daniel M. Evans, C. Andrew Dolloff, W. Michael Aust, and Amy M. Villamagna, "Effects of Eastern Hemlock Decline on Large Wood Loads in Streams of the Appalachian Mountains," *Journal of the American Water Resources Association* 48 (2012): 266–76; Jered M. Studinski et al., "The Effects of Riparian Forest Disturbance on Stream Temperature, Sedimentation, and Morphology," *Hydrobiologia* 686 (2012): 107–17.

10. Chelcy R. Ford and James M. Vose, "*Tsuga canadensis* (L.) Carr: Mortality Will Impact Hydrological Processes in Southern Appalachian Forest Ecosystems," *Ecological Applications* 17 (2007): 1156–67; Brantley et al., "Water Yield and Stormflow."

11. Leigh A. Siderhurst, Heather P. Griscom, Mark Hudy, and Zachary J. Bortolot, "Changes in Light Levels and Stream Temperatures with Loss of Eastern Hemlock (*Tsuga canadensis*) at a Southern Appalachian Stream: Implications for Brook Trout," *Forest Ecology and Management* 260 (2010): 1677–88.

12. Thad E. Yorks, Jennifer C. Jenkins, Donald J. Leopold, Dudley J. Raynal, and Donald A. Orwig, "Influences of Eastern Hemlock Mortality on Nutrient Cycling," in *Proceedings: Sustainable Management of Hemlock Ecosystems in Eastern North America*, ed. Katherine A. McManus, Kathleen S. Shields, and Dennis R. Souto (Durham, NH: US Department of Agriculture, 2000), 126–33.

13. Yorks et al.

14. D. R. Mount et al., "Effect of Long-Term Exposure to Acid, Aluminum, and Low Calcium on Adult Brook Trout (*Salvelinus fontinalis*): 1. Survival, Growth, Fecundity, and Progeny Survival," *Canadian Journal of Fisheries and Aquatic Sciences* 45 (1988): 1623–32.

15. Jennifer Courtwright and Christine L. May, "Importance of Terrestrial Subsidies for Native Brook Trout in Appalachian Intermittent Streams," *Freshwater Biology* 58 (2013): 2423–38.

16. Craig D. Snyder, John A. Young, David P. Lemarié, and David R. Smith, "Influence of Eastern Hemlock (*Tsuga canadensis*) Forests on Aquatic Invertebrate Assemblages in Headwater Streams," *Canadian Journal of Fisheries and Aquatic Sciences* 59 (2002): 262–75; James J. Willacker Jr., William V. Sobczak, and Elizabeth A. Colburn, "Stream Macroinvertebrate Communities in Paired Hemlock and Deciduous Watersheds," *Northeastern Naturalist* 16 (2009): 101–12; Paige M. Kleindl, Fred D. Tucker, Michael G. Commons, Robert G. Verb, and Leslie A. Riley, "Influences of a *Tsuga canadensis* (L.) Carriere (Eastern Hemlock) Riparian Habitat on a Lotic Benthic Community," *Northeastern Naturalist* 23 (2016): 555–70.

17. Jason R. Rohr, Carolyn G. Mahan, and Ke Chung Kim, "Response of Arthropod Biodiversity to Foundation Species Declines: The Case of the Eastern Hemlock," *Forest Ecology and Management* 258 (2009): 1503–10; Rachael E. Mallis and Lynne K. Rieske, "Arboreal Spiders in Eastern Hemlock," *Environmental Entomology* 40 (2011): 1378–87.

18. R. M. Ross, R. M. Bennett, C. D. Snyder, J. A. Young, D. R. Smith, and D. P. Lemarie, "Influence of Eastern Hemlock (*Tsuga canadensis* L.) on Fish Community Structure and Function in Headwater Streams of the Delaware River Basin," *Ecology of Freshwater Fish* 12 (2003): 60–65.

19. Stanley E. Buck III, "Insect Fauna Associated with Eastern Hemlock, *Tsuga canadensis* (L.), in the Great Smoky Mountains National Park" (MS thesis, University of Tennessee, 2004), 47; Carla Dilling, Paris Lambdin, Jerome Grant, and Lee Buck, "Insect Guild Structure Associated with Eastern Hemlock in the Southern Appalachians," *Environmental Entomology* 36 (2007): 1408–14; Carla Coots, Paris Lambdin, Jerome Grant, and Rusty Rhea, "Diversity, Vertical Stratification and Co-occurrence Patterns of the Mycetophilid Community among Eastern Hemlock, *Tsuga canadensis* (L.) Carrière, in the Southern Appalachians," *Forests* 3 (2012): 986–96; Richard M. Turcotte, "Temporal and Spatial Distribution of Imidacloprid and the Arthropod Fauna Associated with Eastern Hemlock, *Tsuga canadensis* (L.) Carr" (PhD diss., West Virginia University, 2016), 49.

20. John A. Sweka and Kyle J. Hartman, "Contribution of Terrestrial Invertebrates to Yearly Brook Trout Prey Consumption and Growth," *Transactions of the American Fisheries Society* 137 (2008): 224–35.

21. Ryan M. Utz, Brett C. Ratcliffe, Brett T. Moore, and Kyle J. Hartman, "Disproportionate Relative Importance of a Terrestrial Beetle Family (Coleoptera: Scarabaeidae) as a Prey Source for Central Appalachian Brook Trout," *Transactions of the American Fisheries Society* 136 (2007): 177–84.

22. L. E. Dodd, Z. Cornett, A. Smith, and L. K. Rieske, "Variation in Lepidopteran Occurrence in Hemlock-Dominated and Deciduous-Dominated Forests of Central Appalachia," *Great Lakes Entomologist* 46 (2003): 1–12.

23. John W. Williams, Bryan N. Shuman, Thompson Webb III, Patrick J. Bartlein, and Phillip L. Leduc, "Late-Quaternary Vegetation Dynamics in North America: Scaling from Taxa to Biomes," *Ecological Monographs* 74 (2004): 309–34.

24. Williams et al.

25. Keith D. Bennett and J. L. Fuller, "Determining the Age of the Mid-Holocene *Tsuga canadensis* (Hemlock) Decline, Eastern North America," *Holocene* 12 (2002): 421–29; Robert K. Booth, Simon Brewer, Maarten Blaauw, Thomas A. Minckley, and Stephen T. Jackson, "Decomposing the Mid-Holocene *Tsuga* Decline in Eastern North America," *Ecology* 93 (2012): 1841–52.

26. Jean Nicolas Haas and John H. McAndrews, "The Summer Drought Related Hemlock (*Tsuga canadensis*) Decline in Eastern North America 5700 to 5100 Years Ago," in *Proceedings: Sustainable Management of Hemlock Ecosystems in Eastern North America,* ed. Katherine A. McManus, Kathleen S. Shields, and Dennis R. Souto (Durham, NH: US Department of Agriculture, 2000), 81–88; Bryan N. Shuman, Paige Newby, and Jeffrey P. Donnelly, "Abrupt Climate Change as an Important Agent of Ecological Change in the Northeast U.S. throughout the Past 15,000 Years," *Quaternary Science Reviews* 28 (2009): 1693–709.

27. Najat Bhiry and Louise Filion, "Mid-Holocene Hemlock Decline in Eastern North America Linked with Phytophagous Insect Activity," *Quaternary Research* 45 (1996): 312–20.

28. Taber D. Allison, Robert E. Moeller, and Margaret Bryan Davis, "Pollen in Laminated Sediments Provides Evidence of Mid-Holocene Forest Pathogen Outbreak," *Ecology* 67 (1986): 1101–5.

29. Amy Hessl and Neil Pederson, "Hemlock Legacy Project (HeLP): A Paleoecological Requiem for Eastern Hemlock," *Progress in Physical Geography* 37 (2012): 114–29.

30. Milutin Milanković, "Mathematische klimalehre und astronomische theorie der klimaschwankungen," in *Handbuch der Klimatologie,* vol. 1, part A, ed. W. Köppen and R. Geiger (Berlin: Borntraeger, 1930), 1–176; Gerard Bond et al., "A Pervasive Millennial-Scale Cycle in North Atlantic Holocene and Glacial Climates," *Science* 278 (1997): 1257–66; Stephen P. Obrochta, Hiroko Miyahara, Yusuke Yokoyama, and Thomas J. Crowley, "A Re-examination of Evidence for the North Atlantic '1500-Year Cycle' at Site 609," *Quaternary Science Reviews* 55 (2012): 23–33.

31. Bryan N. Shuman, Patrick J. Bartlein, and Thompson Webb III, "The Magnitudes of Millennial- and Orbital-Scale Climatic Change in Eastern North America during the Late Quaternary," *Quaternary Science Reviews* 24 (2005): 2194–206.

32. A. E. Viau, K. Gajewski, M. C. Sawada, and P. Fines, "Millennial-Scale Temperature Variations in North America during the Holocene," *Journal of Geophysical Research* 111 (2006): https://doi.org/10.1029/2005JD006031.

33. George H. Denton and Wibjörn Karlén, "Holocene Climatic Variations—Their Pattern and Possible Cause," *Quaternary Research* 3 (1973): 155–205.

34. Brian Menounos, John J. Clague, Gerald Osborn, Brian H. Luckman, Thomas R. Lakeman, and Ryan Minkus, "Western Canadian Glaciers Advance in Concert with Climate Change circa 4.2 ka," *Geophysical Research Letters* 35, no. 7 (2008): https://doi.org/10.1029/2008GL033172; Konrad Gajewski and Andre E. Viau, "Abrupt Climate Changes during the Holocene across North America from Pollen and Paleolimnological Records," in *Abrupt Climate Change: Mechanisms,*

Patterns, and Impacts, ed. Harunur Rashid, Leonid Polyak, and Ellen Mosley-Thompson, Geophysical Monograph Series 193 (Washington, DC: American Geophysical Union, 2011), 173–83; Juzhi Hou, Yongsong Huang, Bryan N. Shuman, W. Wyatt Oswald, and David R. Foster, "Abrupt Cooling Repeatedly Punctuated Early-Holocene Climate in Eastern North America," *Holocene* 22 (2012): 525–29.

35. Bryan N. Shuman and Jeremiah Marsicek, "The Structure of Holocene Climate Change in Mid-Latitude North America," *Quaternary Science Reviews* 141 (2016): 38–51.

36. J. Donald Meisner, "Effect of Climatic Warming on the Southern Margins of the Native Range of Brook Trout, *Salvelinus fontinalis,*" *Canadian Journal of Fisheries and Aquatic Sciences* 47, no. 6 (1990): 1065–70; John G. Eaton and Robert M. Scheller, "Effects of Climate Warming on Fish Thermal Habitat in Streams of the United States," *Limnology and Oceanography* 41 (1996): 1109–15.

37. West Virginia Conservation Commission, *Report of the West Virginia Conservation Commission, 1908* (Charleston, WV: Tribune Printing Company, 1909).

38. West Virginia Conservation Commission.

39. Donald C. Gasper, "West Virginia Trout Streams: Target for Acid Precipitation," in *Confluence 1983,* ed. Rick Webb (Morgantown: West Virginia Mountain Stream Monitors, 1983), 5–11; J. R. Webb, B. J. Cosby, J. N. Galloway, and G. N. Hornberger, "Acidification of Native Brook Trout Streams in Virginia," *Water Resources Research* 25 (1989): 1367–77; Alan T. Herlihy, Philip R. Kaufmann, and Mark E. Mitch, "Regional Estimates of Acid Mine Drainage Impact on Streams in the Mid-Atlantic and Southeastern United States," *Water, Air, and Soil Pollution* 50 (1990): 91–107; David R. DeWalle and Bryan R. Swistock, "Causes of Episodic Acidification in Five Pennsylvania Streams on the Northern Appalachian Plateau," *Water Resources Research* 30 (1994): 1955–63.

40. Thomas F. Waters, "Replacement of Brook Trout by Brown Trout over 15 Years in a Minnesota Stream—Production and Abundance," *Transactions of the American Fisheries Society* 112 (1983): 137–46; Lynn DeWald and Margaret A. Wilzbach, "Interactions between Native Brook Trout and Hatchery Brown Trout—Effects on Habitat Use, Feeding, and Growth," *Transactions of the American Fisheries Society* 121 (1992): 287–96; S. M. Carlson, A. P. Hendry, and B. H. Letcher, "Growth Rate Differences between Resident Native Brook Trout and Non-native Brown Trout," *Journal of Fish Biology* 71 (2007): 1430–47.

41. Scott A. Stranko, Robert H. Hilderbrand, Raymond P. Morgan II, Mark W. Staley, Andrew J. Becker, Ann Roseberry-Lincoln, Elgon S. Perry, and Paul T. Jacobson, "Brook Trout Declines with Land Cover and Temperature Changes in Maryland," *North American Journal of Fisheries Management* 28 (2008): 1223–32.

42. Ira O. Poplar-Jeffers, J. Todd Petty, James T. Anderson, Steven J. Kite, Michael P. Strager, and Ronald H. Fortney, "Culvert Replacement and Stream Habitat Restoration: Implications from Brook Trout Management in an Appalachian Watershed, U.S.A.," *Restoration Ecology* 17 (2009): 404–13.

43. R. Allen Curry and W. Scott MacNeill, "Population-Level Responses to Sediment during Early Life in Brook Trout," *Journal of the North American Benthological Society* 23 (2004): 140–50.

44. Maya Weltman-Fahs and Jason M. Taylor, "Hydraulic Fracturing and Brook Trout Habitat in the Marcellus Shale Region: Potential Impacts and Research Needs," *Fisheries* 38 (2013): 4–15.

45. Mark Hudy, Teresa Thieling, Nathaniel Gillespie, and Eric P. Smith, "Distribution, Status, and Land Use Characteristics of Subwatersheds within the Native Range of Brook Trout in the Eastern United States," *North American Journal of Fisheries Management* 28 (2008): 1069–85.

CHAPTER 13: APPALACHIAN APPOSITION

1. Ole Worm, *Museum wormianum: Seu historia rerum rariorum; Tam naturalium, quam artificialium, tam domesticarum, quam exoticarum, quæ Hafniæ Danorum in ædibus authoris fervantur* (Leiden: Elzevir, 1655), frontispiece.

2. Carl Linnaeus, *Systema naturæ per regna tria naturæ, secundum classes, ordines, genera, species, cum characteribus, differentiis, synonymis, locis,* 10th ed., rev., vol. 1 (Stockholm: Laurentius Salvius, 1758), 7.

3. Zadock Thompson, *History of Vermont, Natural, Civil and Statistical, in Three Parts, with an Appendix* (Burlington, VT: Stacy and Jameson, 1853), 33, 34.

4. Thompson, 34.

5. Louis Agassiz, "On Two New Genera of Fishes from Lake Superior" (presented on November 18, 1848, to 23 members of the Boston Society of Natural History), *Proceedings of the Boston Society of Natural History* 3 (1851): 80–81.

6. Agassiz, "Two New Genera of Fishes," 81.

7. Thompson, *History of Vermont,* 34.

8. Zadock Thompson, "Species Description of *Percopsis pellucida*" (presented to five members of the Boston Society of Natural History on July 18, 1849, by D. H. Storer, on behalf of Zadock Thompson), *Proceedings of the Boston Society of Natural History* 3 (1848–51): 163–65.

9. Thompson, 164.

10. Thompson.

11. Agassiz, "Two New Genera of Fishes."

12. Louis Agassiz, *Lake Superior: Its Physical Character, Vegetation, and Animals, Compared with Those of Other and Similar Regions with a Narrative of the Tour by J. Elliot Cabot, and Contributions by Other Scientific Gentlemen* (Boston: Gould, Kendall and Lincoln, 1850), 286–89.

13. Louis Agassiz, *Recherches sur les Poissons Fossiles,* 4 vols. (Neuchatel, Switzerland: Petitpierre, 1833–44).

14. Agassiz, *Lake Superior,* 24.

15. Agassiz, 285.

16. William J. Poly, *Family Percopsidae Agassiz 1850—Troutperches and Sand Rollers: Annotated Checklist of Fishes, no. 23* (San Francisco: California Academy of Sciences, 2004), 1–5.

17. William C. Kendall, "Notes on *Percopsis guttatus* Agassiz and *Salmo omiscomaycus* Walbaum," *Proceedings of the Biological Society of Washington* 24 (1911): 45–51.

18. Poly, *Troutperches and Sand Rollers,* 2.

19. Thomas Pennant, *Arctic Zoology,* vol. 1 (London: Henry Hughs, 1784), cxcii.

20. W. B. Scott and E. J. Crossman, *Freshwater Fishes of Canada,* Bulletin 184 (Ottawa: Fisheries Research Board of Canada, 1973), 682.

21. Johanne J. Walbaum, *Petri Artedi sueci genera piscium: In quibus systema totum ichthyologiæ proponitur cum classibus, ordinibus, generum characteribus, specierum differentiis, observationibus plurimis; Redactis speciebus 242 ad genera 52; Ichthyologiæ pars III* (Greifswald, Germany: Ferdin, Röse, 1792), 65.

CHAPTER 14: FRAGMENTED THOUGHTS

1. Nathaniel S. Shaler, *Nature and Man in America* (New York: Charles Scribner's Sons, 1897), 9–10.

2. Sally G. Kohlstedt, "Nature, Not Books: Scientists and the Origins of the Nature-Study Movement in the 1890s," *Isis* 96 (2005): 325.

3. Henry W. Longfellow, *The Song of Hiawatha* (Boston: Ticknor and Fields, 1855).

4. Henry W. Longfellow, *The Courtship of Miles Standish, and Other Poems* (Boston: Ticknor and Fields, 1859), 196–98.

5. Robert E. Jenkins and Noel M. Burkhead, *Freshwater Fishes of Virginia* (Bethesda, MD: American Fisheries Society, 1994), 648.

6. Daniel H. Doctor, David J. Weary, David K. Brezinski, Randall C. Orndorff, and Lawrence E. Spangler, "Karst of the Mid-Atlantic Region in Maryland, West Virginia, and Virginia," in *Tripping from the Fall Line: Field Excursions for the GSA Annual Meeting, Baltimore, 2015, Field Guide 40,* ed. David K. Brezinski, Jeffrey P. Halka, and Richard A. Ortt Jr. ([Boulder, CO]: Geological Society of America, 2015), 424–84.

7. Stuart A. Welsh, "The Effects of Long-Term Insularization on Morphological Variation in the Checkered Sculpin" (MS thesis, Frostburg State University, 1996).

8. Nathaniel S. Shaler and Sophia Penn Page Shaler, *The Autobiography of Nathaniel Southgate Shaler with a Supplementary Memoir by His Wife* (New York: Riverside Press; Cambridge, MA: Houghton Mifflin, 1909), 98, 99.

9. Shaler and Shaler, 99–100.

10. Welsh, "Morphological Variation in the Checkered Sculpin."

11. Shaler and Shaler, *Autobiography of Nathaniel Southgate Shaler,* 99.

12. John T. Hack, *Geomorphology of the Shenandoah Valley Virginia and West Virginia and Origin of the Residual Ore Deposits,* US Geological Survey Professional Paper 484 (Washington, DC: US Government Printing Office, 1965).

13. William A. Hobba Jr., Donald W. Fisher, F. J. Pearson Jr., and Joseph C. Chemerys, *Hydrology and Geochemistry of Thermal Springs of the Appalachians,* US Geological Survey Professional Paper 1044-E (Washington, DC: US Government Printing Office, 1979), E22.

14. Jack E. Nolde and William F. Giannini, "Ancient Warm Springs Deposits in Bath and Rockingham Counties, Virginia," *Virginia Minerals* 43 (1997): 9–16.

15. "Thermal Springs in the U.S.," interactive map, National Centers for Environmental Information, accessed February 22, 2023, https://maps.ngdc.noaa.gov/viewers /hot_springs/.

16. Richard M. Yager, L. Neil Plummer, Leon J. Kaufmann, Daniel H. Doctor, David L. Nelms, and Peter Schlosser, "Comparison of Age Distributions Estimated from

Environmental Tracers by Using Binary-Dilution and Numerical Models of Fractured and Folded Karst: Shenandoah Valley of Virginia and West Virginia, USA," *Hydrogeology Journal* 21 (2013): 1193–217.

17. Mark D. Kozar and David J. Weary, *Hydrogeology and Ground-Water Flow in the Opequon Creek Watershed Area, Virginia and West Virginia,* US Geological Survey Scientific Investigations Report 2009-5153 (Reston, VA: US Geological Survey, 2009), 61; Yager et al., "Fractured and Folded Karst."

18. Dorothy J. Vesper, Rachel V. Grand, Kristen Ward, and Joseph J. Donovan, "Geochemistry of a Spring-Dense Karst Watershed Located in a Complex Structural Setting, Appalachian Great Valley, West Virginia, USA," *Environmental Geology* 58 (2009): 667–78.

19. Mark T. Duigon, *Phase 2 Study of the Area Contributing Groundwater to the Spring Supplying the A. M. Powell State Fish Hatchery, Washington County, Maryland,* Maryland Geological Survey, Open-File Report No. 2008-02-18 ([Baltimore]: Maryland Geological Survey, 2008).

20. Larry J. Nutter, "Hydrogeology of Antietam Creek Basin," *Journal of Research, U.S. Geological Survey* 2 (1974): 249–52.

21. David A. Saad and Daniel J. Hippe, *Large Springs in the Valley and Ridge Physiographic Province of Pennsylvania,* US Geological Survey Open-File Report 90-164 (Harrisburg, PA: US Geological Survey, 1990), 1–17.

22. Robert A. Schultz, William A. Hobba Jr., and Mark D. Kozar, *Geohydrology, Ground-Water Availability, and Ground-Water Quality of Berkeley County, West Virginia, with Emphasis on the Carbonate-Rock Area,* Water-Resources Investigations Report 93-4073 (Charleston, WV: US Geological Survey, 1995), 1–88.

23. Yager et al., "Fractured and Folded Karst."

24. Alan R. Templeton, Kerry Shaw, Eric Routman, and Scott K. Davis, "The Genetic Consequences of Habitat Fragmentation," *Annals of the Missouri Botanical Garden* 77 (1990): 13–27.

CHAPTER 15: SATELLITE SUNFISH

1. Vernon L. Avila, "A Field Study of Nesting Behavior of Male Bluegill Sunfish (*Lepomis macrochirus* Rafinesque)," *American Midland Naturalist* 96 (1976): 199.

2. Mart R. Gross, "Evolution of Alternative Reproductive Strategies: Frequency-Dependent Sexual Selection in Male Bluegill Sunfish," *Transactions of the Royal Society of London B* 332 (1991): 59–66.

3. Gross.

4. Sarah E. Magee and Bryan D. Neff, "Temporal Variation in Decisions about Parental Care in Bluegill, *Lepomis macrochirus,*" *Ethology* 112 (2006): 1000–1007.

5. Luca M. Cargnelli and Bryan D. Neff, "Condition-Dependent Nesting in Bluegill Sunfish *Lepomis macrochirus,*" *Journal of Animal Ecology* 75 (2006): 627–33.

6. Corsin F. Muller and Marta B. Manser, "'Nasty Neighbours' Rather than 'Dear Enemies' in a Social Carnivore," *Proceedings of the Royal Society B* 27 (2007): 959–65.

7. Patrick W. Colgan, William A. Nowell, Mart R. Gross, and James W. A. Grant, "Aggressive Habituation and Rim Circling in the Social Organization of Bluegill Sunfish (*Lepomis macrochirus*)," *Environmental Biology of Fishes* 4 (1979): 29–36.

8. Edward O. Wilson, *Sociobiology: The New Synthesis* (Cambridge, MA: Harvard University Press, 1975), 382.

9. James Maxwell McConnell Fisher, "Evolution and Bird Sociality," in *Evolution as a Process*, ed. J. Huxley, A. C. Hardy, and E. B. Ford (London: Allen and Unwin, 1954), 72.

10. Charles M. Breder Jr., "The Reproductive Habits of the North American Sunfishes (Family Centrarchidae)," *Zoologica* 21 (1936): 1–48; William Ingram and Eugene P. Odum, "Nests and Behavior of *Lepomis gibbosus* (Linne) in Lincoln Pond, Rensselaerville, N.Y.," *American Midland Naturalist* 226 (1941): 182–93; John R. Hunter, "The Reproductive Behavior of the Green Sunfish *Lepomis cyanellus*," *Zoologica* 48 (1963): 13–24.

11. Avila, "Nesting Behavior of Male Bluegill," 199.

12. Wallace J. Dominey, "Female Mimicry in Bluegill Sunfish—a Genetic Polymorphism?," *Nature* 284 (1980): 546–48.

13. Mart R. Gross, "Cuckoldry in Sunfishes (*Lepomis*: Centrarchidae)," *Canadian Journal of Zoology* 57 (1979): 1507–9.

14. Gross, "Alternative Reproductive Strategies."

15. Mart R. Gross and Eric L. Charnov, "Alternative Male Life Histories in Bluegill Sunfish," *Proceedings of the National Academy of Sciences of the United States of America* 77 (1980): 6937–40; Gross, "Alternative Reproductive Strategies."

16. Gross, "Alternative Reproductive Strategies."

17. Gross, 60.

18. Gross.

19. Gross, 63.

20. Bryan D. Neff, "Paternity and Condition Affect Cannibalistic Behavior in Nest-Tending Bluegill Sunfish," *Behavioral Ecology and Sociobiology* 54 (2003): 377–84.

21. Bryan D. Neff and Paul W. Sherman, "Nestling Recognition via Direct Cues by Parental Male Bluegill Sunfish (*Lepomis macrochirus*)," *Animal Cognition* 6 (2003): 87–92.

22. Mart R. Gross and Joe Repka, "Stability with Inheritance in the Conditional Strategy," *Journal of Theoretical Biology* 192 (1998): 445–53.

23. John A. Endler, *Natural Selection in the Wild* (Princeton, NJ: Princeton University Press, 1986), 39.

24. Gross and Repka, "Stability with Inheritance," 447.

CHAPTER 16: GAMEST FISH THAT SWIMS

1. James A. Henshall, *Book of the Black Bass Comprising Its Complete Scientific and Life History Together with a Practical Treatise on Angling and Fly Fishing and a Full Description of Tools, Tackle and Implements* (Cincinnati: Robert Clarke, 1881), 380.

2. Bernard G. Lacépède, *Histoire Naturelle des Poissons, IV* (Paris: Chez Plassan, 1802), 323, 716.

3. Henshall, *Book of the Black Bass*, 33.

4. David Starr Jordan and Charles H. Gilbert, *Synopsis of the Fishes of North America,* Bulletin of the United States National Museum, no. 16. (Washington, DC: US Government Printing Office, 1882), 484–85.

5. Henshall, *Book of the Black Bass*, vi.
6. Robert E. Jenkins and Noel M. Burkhead, *Freshwater Fishes of Virginia* (Bethesda, MD: American Fisheries Society, 1994), 729.
7. Henshall, *Book of the Black Bass*, 379.
8. John Eoff, "On the Habits of the Black Bass of the Ohio (*Grystes fasciatus*)," in *Ninth Annual Report of the Board of Regents of the Smithsonian Institution* (Washington, DC: Beverley Tucker, Senate Printer, 1855), 289–90.
9. P. R. Uhler and Otto Lugger, "List of Fishes of Maryland," in *Report of the Commissioners of Fisheries of Maryland to the General Assembly, January 1st 1876* (Annapolis, MD: John F. Wiley, State Printer, 1876), 67–186.
10. Uhler and Lugger.
11. Nancy Clark, "The Hermitage—Legacy of a Century-Old Inn," *Wonderful West Virginia* 43, no. 12 (1980): 2–6.
12. Henshall, *Book of the Black Bass*, 380.
13. James W. Milner, "The Progress of Fish Culture in the United States," in *United States Commission of Fish and Fisheries Report of the Commissioner for 1872 and 1873* (Washington, DC: US Government Printing Office, 1874), 526.
14. Tarleton H. Bean, *The Fishes of Pennsylvania, with Description of the Species and Notes on Their Common Names, Distribution, Habits, Reproduction, Rate of Growth, and Mode of Capture* (Harrisburg, PA: Edwin K. Meyers, Binder, 1892), 117.
15. William E. Meehan, *Fish, Fishing, and Fisheries of Pennsylvania* (Harrisburg, PA: E. K. Meyers, State Printer, 1893), 93.
16. Meehan, 94.
17. Virginia Fish Commission, *Annual Reports of the Fish Commissioners of the State of Virginia for the Years 1875–6 and 1876–7* (Richmond, VA: R. F. Walker, Superintendent Public Printing, 1877), 51.
18. Edward D. Cope, "On the Distribution of Fresh-Water Fishes in the Allegheny Region of Southwestern Virginia," *Journal of the Academy of Natural Sciences of Philadelphia*, 2nd ser., 6, part 3, art. 5 (1869): 207–47.
19. Virginia Fish Commission, *Reports of the Fish Commissioners*, 8.
20. Jenkins and Burkhead, *Freshwater Fishes of Virginia*, 40, 86.
21. Milner, "Fish Culture in the United States," 526.
22. Milner, 526.
23. M. Jake Vander Zanden, Julian D. Olden, James H. Thorne, and Nicholas E. Mandrak, "Predicting Occurrences and Impacts of Smallmouth Bass Introductions in North Temperate Lakes," *Ecological Applications* 14 (2004): 132–48.
24. Michael P. Carey, Beth L. Sanderson, Thomas A. Friesen, Katie A. Barnas, and Julian D. Olden, "Smallmouth Bass in the Pacific Northwest: A Threat to Native Species; A Benefit for Anglers," *Reviews in Fisheries Science* 19 (2011): 305–15.
25. Kei'ichiro Iguchi, Taiga Yodo, and Naoto Matsubara, "Spawning and Brood Defense of Smallmouth Bass under the Process of Invasion into a Novel Habitat," *Environmental Biology of Fishes* 70 (2004): 219–25; Grace L. Loppnow, Kris Vascotto, and Paul A. Venturelli, "Invasive Smallmouth Bass (*Micropterus dolomieu*): History, Impacts, and Control," *Management of Biological Invasions* 4 (2004): 191–206; Darragh J. Woodford, N. Dean Impson, Jenny A. Day, and I. Roger Bills, "The Predatory Impact of Invasive Alien Smallmouth Bass, *Micropterus dolomieu*

(Teleostei: Centrarchidae), on Indigenous Fishes in a Cape Floristic Region Mountain Stream," *African Journal of Aquatic Science* 30, no. 2 (2005): 167–73.

CHAPTER 17: CHARISMATIC CANDY

1. Carl L. Hubbs and Milton B. Trautman, "*Poecilichthys osburni,* a New Darter from the Upper Kanawha River System in Virginia and West Virginia," *Ohio Journal of Science* 32 (1932): 36.
2. C. Richard Robins, Reeve M. Bailey, Carl E. Bond, James R. Booker, Ernest A. Lachner, Robert N. Lea, and W. B. Scott, *Common and Scientific Names of Fishes from the United States and Canada,* 5th ed. (Bethesda, MD: American Fisheries Society, 1991), 49, 89.
3. Robert E. Jenkins and Noel M. Burkhead, *Freshwater Fishes of Virginia* (Bethesda, MD: American Fisheries Society, 1994), 827–30.
4. C. Richard Robins, "*Cottus kanawhae,* a New Cottid Fish from the New River System of Virginia and West Virginia," *Zootaxa* 987 (2005): 1–6; Jenkins and Burkhead, *Freshwater Fishes of Virginia,* 84.
5. Hubbs and Trautman, "*Poecilichthys osburni,* a New Darter," 36.
6. Hubbs and Trautman, 36.
7. John Penczek, Paul A. Boynton, and Jolene D. Splett, "Color Error in the Digital Camera Image Capture Process," *Journal of Digital Imaging* 27 (2014): 182–91.
8. Charles Darwin, *On the Origin of Species by Means of Natural Selection, or the Preservation of Favoured Races in the Struggle for Life* (London: John Murray, 1859), 77–178.
9. Charles Darwin, *The Descent of Man, and Selection in Relation to Sex* (London: John Murray, 1871), 256.
10. Malte Andersson, *Sexual Selection* (Princeton, NJ: Princeton University Press, 1994), 3.
11. Thierry Hoquet and Michael Levandowsky, "Utility vs Beauty: Darwin, Wallace and the Subsequent History of the Debate on Sexual Selection," in *Current Perspectives on Sexual Selection,* ed. Thierry Hoquet (Dordrecht, Netherlands: Springer, 2015), 19–44.
12. Jared P. Kirtland, "Descriptions of Four New Species of Fishes," *Boston Journal of Natural History* 3 (1840): 275.
13. Isaac Gibson, Amy B. Welsh, Stuart A. Welsh, and Daniel A. Cincotta, "Genetic Swamping and Possible Species Collapse: Tracking Introgression between the Native Candy Darter and Introduced Variegate Darter," *Conservation Genetics* 20 (2019): 287–98.
14. Daniel A. Cincotta, Douglas B. Chambers, and Terence Messinger, "Recent Changes in the Distribution of Fish Species in the New River Basin in West Virginia and Virginia," in *Proceedings of the New River Symposium, April 15–16, Boone, North Carolina* (Glen Jean, WV: US National Park Service, 1999), 98–106.
15. David I. Wellman, "Post-flood Recovery and Distributions of Fishes in the New River Gorge National River and the Gauley River National Recreation Area" (MS thesis, West Virginia University, 2004), 1–169.
16. Isaac Gibson, "Conservation Concerns for the Candy Darter (*Etheostoma osburni*) with Implications Related to Hybridization" (MS thesis, West Virginia

University, 2017); John F. Switzer, Stuart A. Welsh, and Tim L. King, "A Molecular Genetic Investigation of Hybridization Between *Etheostoma osburni* and *Etheostoma variatum* in the New River Drainage, West Virginia" (report submitted to the West Virginia Division of Natural Resources, Elkins, West Virginia, 2007); John F. Switzer, Stuart A. Welsh, and Tim L. King, "Microsatellite DNA Primers for the Candy Darter, *Etheostoma osburni* and Variegate Darter, *Etheostoma variatum*, and Cross-Species Amplification in Other Darters (Percidae)," *Molecular Ecology Resources* 8 (2008): 335–38.

17. US Fish and Wildlife Service, "Endangered and Threatened Wildlife and Plants: Designation of Critical Habitat for the Candy Darter," *Federal Register* 83 (2018): 59232–68; US Fish and Wildlife Service, "Endangered and Threatened Wildlife and Plants: Endangered Species Status for the Candy Darter," *Federal Register* 83 (2018): 58747–54.

CHAPTER 18: LIKE A ROLLING STONE

1. Richard Meryman, "A Painter of Angels Became the Father of Camouflage," *Smithsonian* 30, no. 1 (1999): 116–29; Roy R. Behrens, "Revisiting Abbott Thayer: Non-scientific Reflections about Camouflage in Art, War and Zoology," *Philosophical Transactions of the Royal Society of London B: Biological Sciences* 364 (2009): 497–501.
2. Abbott H. Thayer, "Protective Coloration in Its Relation to Mimicry, Common Warming Colours, and Sexual Selection," *Transactions of the Entomological Society of London* 51 (1903): 554.
3. Abbott H. Thayer, *Concealing-Coloration in the Animal Kingdom: An Exposition of the Laws of Disguise through Color and Pattern: Being a Summary of Abbott H. Thayer's Discoveries* (New York: Macmillan, 1909), 80.
4. Theodore Roosevelt, "Revealing and Concealing Coloration in Birds and Mammals," *Bulletin of the American Museum of Natural History* 30 (1911): 120–221.
5. Stephen J. Gould, "Red Wings in the Sunset," *Natural History* 94 (1985): 12–24.
6. Robert Hendrick, "A Raven in a Coal Scuttle: Theodore Roosevelt and the Animal Coloration Controversy," *American Biology Teacher* 57 (1995): 14–20.
7. Graeme D. Ruxton, Michael P. Speed, and David J. Kelly, "What, if Anything, Is the Adaptive Function of Countershading?," *Animal Behaviour* 68 (2004): 445–51.
8. Abbott H. Thayer, "The Law Which Underlies Protective Coloration," *Auk* 13 (1896): 124.
9. Sharon Kingsland, "Abbott Thayer and the Protective Coloration Debate," *Journal of the History of Biology* 11 (1978): 223–44.
10. Martin Stevens, Innes C. Cuthill, C. Alejandro Parraga, and Tom Troscianko, "The Effectiveness of Disruptive Coloration as a Concealment Strategy," *Progress in Brain Research* 155 (2006): 49–64.
11. H. Martin Schaefer and Nina Stobbe, "Disruptive Coloration Provides Camouflage Independent of Background Matching," *Proceedings of the Royal Society B* 273 (2006): 2427–32.
12. F. W. Campbell and L. Maffei, "Contrast and Spatial Frequency," *Scientific American* 231, no. 5 (1974): 106–15.
13. Campbell and Maffei.

14. Martin Stevens, W. Tom L. Searle, Jenny E. Seymour, Kate L. Marshall, and Graeme D. Ruxton, "Motion Dazzle and Camouflage as Distinct Anti-predator Defenses," *BMC Biology* 9 (2011): 81; Bettina von Helversen, Lael J. Schooler, and Uwe Czienskowski, "Are Stripes Beneficial? Dazzle Camouflage Influences Perceived Speed and Hit Rates," *PLoS ONE* 8, no. 4 (2013): e61173, https://doi.org /10.1371/journal.pone.0061173; Anna E. Hughes, Jolyon Troscianko, and Martin Stevens, "Motion Dazzle and the Effects of Target Patterning on Capture Success," *BMC Evolutionary Biology* 14 (2014): 201.

15. F. W. Campbell and L. Maffei, "The Influence of Spatial Frequency and Contrast on the Perception of Moving Patterns," *Vision Research* 21 (1981): 713–21; Pooya Pakarian and Mohammad Taghi Yasamy, "Wagon-Wheel Illusion under Steady Illumination: Real or Illusory?," *Perception* 32 (2003): 1307–10.

16. Sophie Wuerger, Robert Shapley, and Nava Rubin, "'On the Visually Perceived Direction of Motion' by Hans Wallach: 60 Years Later," *Perception* 25 (1996): 1317–67; Martin J. How and Johannes M. Zanker, "Motion Camouflage Induced by Zebra Stripes," *Zoology* 117 (2014): 163–70; Benedict G. Hogan, Innes C. Cuthill, and Nicholas E. Scott-Samuel, "Dazzle Camouflage and the Confusion Effect: The Influence of Varying Speed on Target Tracking," *Animal Behaviour* 123 (2017): 349–53.

17. Peter Thompson, "Does My Butt Look Big in This? Horizontal Stripes, Perceived Body Size and the Oppel-Kundt Illusion," *Journal of Vision* 8 (2008): 822; Peter Thompson and Kyriaki Mikellidou, "The 3-D Helmholtz Square Illusion: More Reasons to Wear Horizontal Stripes," *Journal of Vision* 9 (2009): 50; Peter Thompson and Kyriaki Mikellidou, "Applying the Helmholtz Illusion to Fashion: Horizontal Stripes Won't Make You Look Fatter," *i-Perception* 2 (2011): 69–76.

18. Graeme D. Ruxton, "The Possible Fitness Benefits of Striped Coat Coloration for Zebra," *Mammal Review* 32 (2002): 237–44.

19. James M. Brown and Naomi Weisstein, "A Spatial Frequency Effect on Perceived Depth," *Perception and Psychophysics* 44 (1988): 157–66.

20. Martin Stevens, "The Role of Eyespots as Anti-predator Mechanisms, Principally Demonstrated in the Lepidoptera," *Biological Reviews* 80 (2005): 573–88.

21. Martin Stevens, Elinor Hopkins, William Hinde, Amabel Adcock, Yvonne Connolly, Tom Troscianko, and Innes C. Cuthill, "Field Experiments on the Effectiveness of 'Eyespots' as Predator Deterrents," *Animal Behaviour* 74 (2007): 1215–27; Martin Stevens, Chloe J. Hardman, and Claire L. Stubbins, "Conspicuousness, Not Eye Mimicry, Makes 'Eyespots' Effective Antipredator Signals," *Behavioral Ecology* 19 (2008): 525–31; Martin Stevens and Graeme D. Ruxton, "Do Animal Eyespots Really Mimic Eyes?," *Current Zoology* 60 (2014): 26–36.

CHAPTER 19: TIT-FOR-TAT

1. Andrew Sih, "Optimal Behavior: Can Foragers Balance Two Conflicting Demands?," *Science* 210 (1980): 1041–43.

2. T. J. Pitcher, D. A. Green, and A. E. Magurran, "Dicing with Death: Predator Inspection Behaviour in Minnow Shoals," *Journal of Fish Biology* 28 (1986): 438–48.

3. L. A. Dugatkin, "Do Guppies Play TIT FOR TAT during Predator Inspection Visits?," *Behavioral Ecology and Sociobiology* 25 (1988): 395–99.

4. A. E. Magurran and T. J. Pitcher, "Provenance, Shoal Size and the Sociobiology of Predator Evasive Behaviour in Minnow Shoals," *Proceedings of the Royal Society* 229B (1987): 439–65; K. E. Murphy and T. J. Pitcher, "Individual Behavioural Strategies Associated with Predator Inspection in Minnow Shoals," *Ethology* 88 (1991): 307–19; Lee A. Dugatkin and Jean-Guy J. Godin, "Predator Inspection, Shoaling and Foraging under Predation Hazard in the Trinidadian Guppy, *Poecilia reticulata*," *Environmental Biology of Fishes* 34 (1992): 265–76.

5. Robert Axelrod and William D. Hamilton, "The Evolution of Cooperation," *Science* 211 (1981): 1390–96; Robert Axelrod, *The Evolution of Cooperation* (New York: Basic Books, 1984), 217.

6. J. Maynard Smith and G. R. Price, "The Logic of Animal Conflict," *Nature* 246 (1973): 15–18; J. Maynard Smith, *Evolution and the Theory of Games* (Cambridge: Cambridge University Press, 1982).

7. Josef Hofbauer and Karl Sigmund, *Evolutionary Games and Populations Dynamics* (Cambridge: Cambridge University Press, 1988), 191–97.

8. Axelrod and Hamilton, "Evolution of Cooperation."

9. Manfred Milinski, "Tit for Tat in Sticklebacks and the Evolution of Cooperation," *Nature* 325 (1987): 433–35.

10. Axelrod and Hamilton, "Evolution of Cooperation," 1393.

11. Jean-Guy J. Godin and Shelley L. Crossman, "Hunger-Dependent Predator Inspection and Foraging Behaviours in the Threespine Stickleback (*Gasterosteus aculeatus*) under Predation Risk," *Behavioral Ecology and Sociobiology* 34 (1994): 359.

12. Manfred Milinski, David Kulling, and Rolf Kettler, "Tit for Tat: Sticklebacks 'Trusting' a Cooperating Partner," *Behavioral Ecology* 1 (1990): 7–11.

13. Manfred Milinski, "Risk of Predation of Parasitized Sticklebacks (*Gasterosteus aculeatus* L.) under Competition for Food," *Behaviour* 93 (1985): 203–16; Pitcher, Green, and Magurran, "Behaviour in Minnow Shoals."

14. Thomas Licht, "Discriminating between Hungry and Satiated Predators: The Response of Guppies (*Poecilia reticulata*) from High and Low Predation Sites," *Ethology* 82 (1989): 238–43; Anne E. Magurran and Sarah L. Girling, "Predator Model Recognition and Response Habituation in Shoaling Minnows," *Animal Behaviour* 34 (1986): 510–18; Anne E. Magurran and Anthony Higham, "Information Transfer across Fish Shoals under Predator Threat," *Ethology* 78 (1988): 153–58.

15. Anne E. Magurran, "Individual Differences in Fish Behaviour," in *The Behaviour of Teleost Fishes,* ed. T. J. Pitcher (London: Croom Helm, 1986), 338–65; K. E. Murphy and T. J. Pitcher, "Predator Attack Motivation Influences the Inspection Behaviour of European Minnows," *Journal of Fish Biology* 50 (1997): 407–17.

16. Pitcher, Green, and Magurran, "Behavior in Minnow Shoals," 447.

17. Axelrod and Hamilton, "Evolution of Cooperation," 1391.

CHAPTER 20: DISCOVERING DIAMONDS

1. Stuart A. Welsh and Robert M. Wood, "*Crystallaria cincotta,* a New Species of Darter (Teleostei: Percidae) from the Elk River of the Ohio River Drainage, West Virginia," *Zootaxa* 1680 (2008): 62–68.

2. R. E. Call, *The Life and Writings of Rafinesque* (Louisville, KY: John P. Morton, 1895); T. J. Fitzpatrick, *Rafinesque: A Sketch of His Life with Bibliography* (Des Moines: Historical Department of Iowa, 1911).

3. Constantine S. Rafinesque, *Ichthyologia Ohiensis or Natural History of the Stream Fishes Inhabiting the River Ohio and Its Natural Tributary Streams* (Lexington, KY: W. G. Hunt, 1820), 5.

4. David Starr Jordan, "Rafinesque," *Popular Science Monthly* 29 (1886): 218.

5. C. Boewe, "The Fall from Grace of That 'Base Wretch' Rafinesque," *Kentucky Review* 7 (1987): 39–53; A. Wheeler, "An Appraisal of the Zoology of C. S. Rafinesque," *Bulletin of Zoological Nomenclature* 45 (1988): 6–12; David Starr Jordan, "Concerning the Fishes of the Ichthyologia Ohiensis," *Bulletin of the Buffalo Society of Natural Sciences* 3 (1877): 91–97.

6. Lawrence M. Page, Héctor Espinosa-Pérez, Lloyd T. Findley, Carter R. Gilbert, Robert N. Lea, Nicholas E. Mandrak, Richard L. Mayden, and Joseph S. Nelson, *Common and Scientific Names of Fishes from the United States, Canada, and Mexico, Special Publication 34* (Bethesda, MD: American Fisheries Society, 2013).

7. Douglas F. Markle, "Audubon's Hoax: Ohio River Fishes Described by Rafinesque," *Archives of Natural History* 24 (1997): 439–47; Neal Woodman, "Pranked by Audubon: Constantine S. Rafinesque's Description of John James Audubon's Imaginary Kentucky Mammals," *Archives of Natural History* 43 (2016): 95–108.

8. Francis H. Herrick, *Audubon the Naturalist: A History of His Life and Time,* vol. 1 (New York: D. Appleton, 1917).

9. John J. Audubon, *Ornithological Biography,* vol. 1. (Philadelphia: E. L. Carey and A. Hart, 1832), 455–56.

10. Audubon, 456.

11. Audubon, 457.

12. Markle, "Audubon's Hoax."

13. David Starr Jordan, *Review of Rafinesque's Memoirs on North American Fishes,* Bulletin of the United States National Museum, no. 9 (Washington, DC: US Government Printing Office, 1877).

14. Daniel A. Cincotta and Michael E. Hoeft, "Rediscovery of the Crystal Darter, *Ammocrypta asprella,* in the Ohio River Basin," *Brimleyana* 13 (1987): 133–36.

15. David Starr Jordan, "A Catalogue of the Fishes of Illinois," *Bulletin of the Illinois State Laboratory of Natural History* 1 (1878): 38–70.

16. Stuart A. Welsh, Dustin M. Smith, and Nate D. Taylor, "Microhabitat Use of the Diamond Darter," *Ecology of Freshwater Fish* 22 (2013): 587–95.

17. Crystal L. Ruble, "Captive Propagation, Reproductive Biology, and Early Life History of *Crystallaria cincotta* (Diamond Darter), *Etheostoma wapiti* (Boulder Darter), *E. vulneratum* (Wounded Darter), and *E. maculatum* (Spotted Darter)" (MS thesis, West Virginia University, 2013); Crystal L. Ruble, Patrick L. Rakes, John R. Shute, and Stuart A. Welsh, "Captive Propagation, Reproductive Biology, and Early Life History of the Diamond Darter (*Crystallaria cincotta*)," *American Midland Naturalist* 172 (2014): 107–18.

18. Conservation Fisheries, "Diamond Darters Spawning at CFI 2010," accessed July 10, 2019, https://www.youtube.com/watch?v=tQSvqXX_az8.

19. Ruble et al., "Captive Propagation," 111.

20. US Fish and Wildlife Service, "Endangered and Threatened Wildlife and Plants: Endangered Species Status for Diamond Darter; Final Rule," *Federal Register* 78 (2013): 45074–95.

CHAPTER 21: FISH PHYSICS

1. Osborne Reynolds, *On an Inversion of Ideas as to the Structure of the Universe*, Rede Lecture, June 10, 1902 (Cambridge: Cambridge University Press, 1903).
2. Osborne Reynolds, *Papers on Mechanical and Physical Subjects*, vol. 3, *The Submechanics of the Universe* (Cambridge: Cambridge University Press, 1903).
3. Osborne Reynolds, "On the Dilatancy of Media Composed of Rigid Particles in Contact: With Experimental Illustrations," *London, Edinburgh, and Dublin Philosophical Magazine and Journal of Science* 20 (1885): 470.
4. Osborne Reynolds, "Experiments Showing Dilatancy, a Property of Granular Material, Possibly Connected with Gravitation," in *Notices of the Proceedings at the Meeting of the Members of the Royal Institution of Great Britain with Abstracts of the Discourses Delivered at the Evening Meetings* (London: William Clowes and Sons, 1887), 11:360.
5. Daniel M. Evans and Lawrence M. Page, "Distribution and Relative Size of the Swim Bladder in *Percina*, with Comparisons to *Etheostoma, Crystallaria*, and *Ammocrypta* (Teleostei: Percidae)," *Environmental Biology of Fishes* 66 (2003): 61–65.
6. Robert A. Daniels, "Habitat of the Eastern Sand Darter, *Ammocrypta pellucida*," *Journal of Freshwater Ecology* 8 (1993): 287–95; Shannon M. O'Brien and Douglas E. Facey, "Habitat Use by the Eastern Sand Darter, *Ammocrypta pellucida*, in Two Lake Champlain Tributaries," *Canadian Field-Naturalist* 122 (2008): 239–46; Patricia A. Thompson, Stuart A. Welsh, Austin A. Rizzo, and Dustin M. Smith, "Effect of Substrate Size on Sympatric Sand Darter Benthic Habitat Preferences," *Journal of Freshwater Ecology* 32 (2017): 455–65.
7. Patricia A. Thompson, Stuart A. Welsh, Michael P. Strager, and Austin A. Rizzo, "A Multiscale Investigation of Habitat Use and Within-River Distribution of Sympatric Sand Darter Species," *Journal of Geospatial Applications in Natural Resources* 2 (2018). https://scholarworks.sfasu.edu/j_of_geospatial_applications_in_natural_resources/vol2/iss1/1.
8. Paul W. Webb, *Hydrodynamics and Energetics of Fish Propulsion*, Bulletin 190 of the Fisheries Research Board of Canada (Ottawa: Department of the Environment, Fisheries and Marine Service, 1975).
9. John G. Lundberg and Edie Marsh, "Evolution and Functional Anatomy of the Pectoral Fin Rays in Cyprinoid Fishes: With Emphasis on the Suckers (Family Catostomidae)," *American Midland Naturalist* 96 (1976): 332–49; W. J. Matthews, "Critical Current Speeds and Microhabitats of the Benthic Fishes *Percina roanoka* and *Etheostoma flabellare*," *Environmental Biology of Fishes* 12 (1985): 303–8; Paul W. Webb, Cynthia L. Gerstner, and Scott T. Minton, "Station-Holding by the Mottled Sculpin, *Cottus bairdi* (Teleostei: Cottidae), and Other Fishes," *Copeia* 1996 (1996): 488–93.
10. Thompson et al., "Sympatric Sand Darter Species."
11. P. A. Carling, "The Nature of the Fluid Boundary Layer and the Selection of Parameters for Benthic Ecology," *Freshwater Biology* 28 (1992): 273–84.

12. Osborne Reynolds, "An Experimental Investigation of the Circumstances Which Determine Whether the Motion of Water Shall Be Direct or Sinuous, and of the Law of Resistance in Parallel Channels," *Philosophical Transactions of the Royal Society A* 174 (1883): 935–82.

13. Francois G. Schmitt, "Turbulence from 1870 to 1920: The Birth of a Noun and of a Concept," *Comptes Rendus Mécanique* 345 (2017): 621.

14. Ludwig Prandtl, "Über flüssigkeitsbewegung bei sehr kleiner reibung," in *Verhandlungen des III: Internationalen Mathematiker-Kongresses in Heidelberg* (Leipzig: B. G. Teubner, 1905), 484–91.

15. John D. Anderson Jr., "Ludwig Prandtl's Boundary Layer," *Physics Today* 58 (2005): 42–48.

16. Prandtl, "Über flüssigkeitsbewegung," 487; translated.

17. Prandtl.

18. Robert W. Blake, "Biomechanics of Rheotaxis in Six Teleost Genera," *Canadian Journal of Zoology* 84 (2006): 1173–86.

19. Anderson, "Prandtl's Boundary Layer."

20. Brian Dean and Bharat Bhushan, "Shark-Skin Surfaces for Fluid-Drag Reduction in Turbulent Flow: A Review," *Philosophical Transactions of the Royal Society of London A* 368 (2010): 4775–806.

21. Dylan K. Wainwright and George V. Lauder, "Mucus Matters: The Slippery and Complex Surfaces of Fish," in *Functional Surfaces in Biology III,* ed. Stanislav N. Gorb and Elena V. Gorb (Cham, Switzerland: Springer, 2017), 223–46.

22. August E. Spreitzer, "Life History, External Morphology, Osteology of the Eastern Sand Darter *Ammocrypta pellucida* (Putnam, 1863), an Endangered Ohio Species (Pisces: Percidae)" (MS thesis, Ohio State University, 1979).

23. Dylan K. Wainwright and George V. Lauder, "Three-Dimensional Analysis of Scale Morphology in Bluegill Sunfish, *Lepomis macrochirus,*" *Zoology* 119 (2016): 182–95.

24. Jin Choi, Woo-Pyung Jeon, and Haecheon Choi, "Mechanism of Drag Reduction by Dimples on a Sphere," *Physics of Fluids* 18 (2006): 041702, https://doi.org/10.1063/1.2191848; Dean and Bhushan, "Shark-Skin Surfaces."

25. George V. Lauder, Dylan K. Wainwright, August G. Domel, James C. Weaver, Li Wen, and Katia Bertoldi, "Structure, Biomimetics, and Fluid Dynamics of Fish Skin Surfaces," *Physical Review Fluids* 1 (2016): 060502, https://doi.org/10.1103/PhysRevFluids.1.060502.

26. Rose L. Carlson and George V. Lauder, "Living on the Bottom: Kinematics of Benthic Station-Holding in Darter Fishes (Percidae: Etheostomatinae)," *Journal of Morphology* 271 (2010): 25–35.

27. Lundberg and Marsh, "Pectoral Fin Rays"; Matthews, "Critical Current Speeds."

28. Lundberg and Marsh, "Pectoral Fin Rays."

29. Sheryl Coombs, Erik Anderson, Christopher B. Braun, and Mark Grosenbaugh, "The Hydrodynamic Footprint of a Benthic, Sedentary Fish in Unidirectional Flow," *Journal of the Acoustical Society of America* 122 (2007): 1227–37; Rose L. Carlson and George V. Lauder, "Escaping the Flow: Boundary Layer Use by the Darter *Etheostoma tetrazonum* (Percidae) during Benthic Station Holding," *Journal of Experimental Biology* 214 (2011): 1181–93.

30. Matthews, "Critical Current Speeds"; Carlson and Lauder, "Benthic Station-Holding."

31. David Starr Jordan and Herbert E. Copeland, "The Sand Darter," *American Naturalist* 11 (1877): 87–88.

32. Robert A. Daniels, "Significance of Burying in *Ammocrypta pellucida*," *Copeia* (1989): 29–34.

33. Spreitzer, "Eastern Sand Darter."

34. Jordan and Copeland, "Sand Darter."

35. Spreitzer, "Eastern Sand Darter."

36. Milton B. Trautman, *The Fishes of Ohio* (Columbus: Ohio State University Press, 1981), 649.

37. Thomas P. Simon, "Startle Response and Causes of Burying Behavior in Captive Eastern Sand Darters, *Ammocrypta pellucida* (Putnam)," *Proceedings of the Indiana Academy of Science* 100 (1991): 155–60.

38. Schmitt, "Turbulence"; N. Rott, "Note on the History of the Reynolds Number," *Annual Review of Fluid Mechanics* 22 (1990): 1–11; Pierre-Yves Lagrée, "Boundary Layer Separation and Asymptotics from 1904 to 1969," *Comptes Rendus Mecanique* 345 (2017): 613–19.

39. Kelly M. Dorgan, Peter A. Jumars, Bruce D. Johnson, and Bernard P. Boudreau, "Macrofaunal Burrowing: The Medium Is the Message," *Oceanography and Marine Biology: An Annual Review* 44 (2006): 85–121.

CHAPTER 22: TIPPECANOE IS TINY TOO

1. DeWitt Clinton, "Description of a New Species of Fish from the Hudson River (*Clupea hudsonia*)," *Annals of the Lyceum of Natural History of New York* 1 (1824): 49–50.

2. Constantine S. Rafinesque, "Descriptions of Two New Genera of North American Fishes, *Opsanus* and *Notropis*," *American Monthly Magazine and Critical Review* 2 (1818): 203–4.

3. Constantine S. Rafinesque, "On 3 N. Sp. of *Clintonia*," *Atlantic Journal* 1 (1832): 120.

4. Lawrence M. Page and Brooks M. Burr, *Peterson Field Guide to Freshwater Fishes of North America North of Mexico,* 2nd ed. (New York: Houghton Mifflin Harcourt, 2011), 508.

5. Neil H. Douglas, "A New Record Size for Darters," *Proceedings of the Louisiana Academy of Sciences* 31 (1968): 41–42.

6. Micah G. Bennett and Kevin W. Conway, "An Overview of North America's Diminutive Freshwater Fish Fauna," *Ichthyological Exploration of Freshwaters* 21 (2010): 63–72.

7. Lawrence M. Page and Brooks M. Burr, "The Smallest Species of Darter (Pisces: Percidae)," *American Midland Naturalist* 101 (1979): 452–53.

8. David Starr Jordan and Barton W. Evermann, "Description of a New Species of Fish from Tippecanoe River, Indiana," *Proceedings of the United States National Museum* 13 (1890): 3–4.

9. James Hankin and David B. Wake, "Miniaturization of Body Size: Organismal Consequences and Evolutionary Significance," *Annual Review of Ecology and Systematics* 24 (1993): 501–19.

10. Stuart A. Welsh and Sue A. Perry, "Habitat Partitioning in a Community of Darters in the Elk River, West Virginia," *Environmental Biology of Fishes* 51 (1998): 411–19.

11. E. van Snik Gray, J. M. Boltz, K. A. Kellogg, and J. R. Stauffer Jr., "Food Resource Partitioning by Nine Sympatric Darter Species," *Transactions of the American Fisheries Society* 126 (1997): 822–40.

12. Melvin L. Warren Jr., Brooks M. Burr, and Bernard R. Kuhajda, "Aspects of the Reproductive Biology of *Etheostoma tippecanoe* with Comments on Egg-Burying Behavior," *American Midland Naturalist* 116 (1986): 215–18.

13. US Fish and Wildlife Service, *Species Status Assessment (SSA) Report for the Tippecanoe Darter (*Etheostoma tippecanoe*), Version 1.0* (Hadley, MA: US Fish and Wildlife Service, 2018), 18.

CHAPTER 23: DARTER OF DARTERS

1. David Starr Jordan, *The Days of a Man: Being Memories of a Naturalist, Teacher, and Minor Prophet of Democracy* (Yonkers-on-Hudson, NY: World Book, 1922), 1:vii.

2. Carl L. Hubbs, "History of Ichthyology in the United States after 1850," *Copeia* 1964 (1964): 42–60.

3. Anonymous, "Literary Notices: Science Sketches, by David Starr Jordan," *Popular Science Monthly* 32 (1888): 563.

4. David Starr Jordan, *Science Sketches* (Chicago: A. C. McClurg, 1887), 30–31.

5. Jordan, 31.

6. Robert E. Jenkins and Noel M. Burkhead, *Freshwater Fishes of Virginia* (Bethesda, MD: American Fisheries Society, 1994), 875.

7. Howard E. Winn, "Comparative Reproductive Behavior and Ecology of Fourteen Species of Darters (Pisces-Percidae)," *Ecological Monographs* 28 (1958): 155–91.

8. Frank B. Cross, *Handbook of Fishes of Kansas,* Miscellaneous Publications of the Museum of Natural History, University of Kansas 45 (Lawrence, KS: Museum of Natural History, 1967).

9. Jordan, *Science Sketches,* 31.

10. Jordan, 31.

11. Lawrence M. Page and David L. Swofford, "Morphological Correlates of Ecological Specialization in Darters," *Environmental Biology of Fishes* 2 (1984): 139–59; Lawrence M. Page and Henry L. Bart Jr., "Egg-Mimics in Darters (Pisces: Percidae)," *Copeia* 1989 (1989): 514–17.

12. Winn, "Ecology of Fourteen Species of Darters," 177–78.

13. Charles T. Lake, "The Life History of the Fan-Tailed Darter *Catonotus flabellaris flabellaris* (Rafinesque)," *American Midland Naturalist* 17 (1936): 816–30.

14. Winn, "Ecology of Fourteen Species of Darters."

15. Lake, "Life History of the Fan-Tailed Darter."

16. Jason H. Knouft, Lawrence M. Page, and Michael J. Plewa, "Antimicrobial Egg Cleaning by the Fringed Darter (Perciformes: Percidae: *Etheostoma crossopterum*): Implications of a Novel Component of Parental Care in Fishes," *Proceedings of the Royal Society of London B* 270 (2003): 2405–11.

17. Henry L. Bart Jr. and Lawrence M. Page, "Morphology and Adaptive Significance of Fin Knobs in Egg-Clustering Darters," *Copeia* 1991 (1991): 80–86.

18. Lake, "Life History of the Fan-Tailed Darter."

19. Kai Lindstrom and R. Craig Sargent, "Food Access, Brood Size and Filial Cannibalism in the Fantail Darter, *Etheostoma flabellare*," *Behavioral Ecology and Sociobiology* 40 (1997): 107–10.

20. Sievert Rowher, "Parent Cannibalism of Offspring and Egg Raiding as a Courtship Strategy," *American Naturalist* 112 (1978): 429–40.

21. Lindstrom and Sargent, "Filial Cannibalism in the Fantail Darter."

22. Roland A. Knapp and Robert Craig Sargent, "Egg-Mimicry as a Mating Strategy in the Fantail Darter: Females Prefer Males with Eggs," *Behavioral Ecology and Sociobiology* 25 (1989): 321–26.

23. Stephen Pruett-Jones, "Independent versus Nonindependent Mate Choice: Do Females Copy Each Other?," *American Naturalist* 140 (1992): 1000–1009.

24. R. Craig Sargent, "Paternal Care and Egg Survival Both Increase with Clutch Size in the Fathead Minnow, *Pimephales promelas*," *Behavioral Ecology and Sociobiology* 23 (1988): 33–37.

25. Sarah B. M. Kraak and Ton G. G. Groothuis, "Female Preference for Nests with Eggs Is Based on the Presence of the Eggs Themselves," *Behaviour* 131 (1994): 189–206.

26. Knapp and Sargent, "Egg-Mimicry as a Mating Strategy."

27. Knouft, Page, and Plewa, "Antimicrobial Egg Cleaning."

APPENDIX 2

1. Charles W. Smiley, "Notes on the Edible Qualities of German Carp and Hints about Cooking Them," *Bulletin of the United States Fish Commission* 3 (1883): 305–32.

BIBLIOGRAPHY

Abbott, Charles C. "Descriptions of New Species of American Fresh-Water Fishes." *Proceedings of the Academy of Natural Sciences of Philadelphia* 12 (1860): 325.

———. *In Nature's Realm.* Trenton, NJ: Albert Brandt, 1900.

———. *Notes of the Night and Other Outdoor Sketches.* New York: Century, 1896.

———. *Outings at Odd Times.* New York: D. Appleton, 1890.

———. *The Rambles of an Idler.* Philadelphia: George W. Jacobs, 1906.

Abella, Scott R. "Impacts and Management of Hemlock Woolly Adelgid in National Parks of the Eastern United States." *Southeastern Naturalist* 13 (2014): 16–45.

Agassiz, Alexander. "The Development of *Lepidosteus,* Part I." *Proceedings of the American Academy of Arts and Sciences* 14 (1878): 65–76.

Agassiz, Louis. *Lake Superior: Its Physical Character, Vegetation, and Animals, Compared with Those of Other and Similar Regions with a Narrative of the Tour by J. Elliot Cabot, and Contributions by Other Scientific Gentlemen.* Boston: Gould, Kendall and Lincoln, 1850.

———. "On Two New Genera of Fishes from Lake Superior." Presented on November 18, 1848, to 23 members of the Boston Society of Natural History. *Proceedings of the Boston Society of Natural History* 3 (1848–51): 80–81.

———. *Recherches sur les poissons fossiles.* Neuchatel, Switzerland: Imprimerie de Petitpierre, 1833–44.

Aldinger, Joni L., and Stuart A. Welsh. "Diel Periodicity and Chronology of Upstream Migration in Yellow-Phase American Eels (*Anguilla rostrata*)." *Environmental Biology of Fishes* 100 (2017): 829–38.

Alexander, Nancy J. "Comparison of a and b Keratin in Reptiles." *Zeitschrift für Zellforschung und mikroskopische Anatomie* 110 (1970): 153–65.

Allison, Taber D., Robert E. Moeller, and Margaret Bryan Davis. "Pollen in Laminated Sediments Provides Evidence of Mid-Holocene Forest Pathogen Outbreak." *Ecology* 67 (1986): 1101–5.

American Veterinary Medical Association. "U.S. Pet Ownership Statistics." Accessed July 10, 2021. https://www.avma.org/resources-tools/reports-statistics/us-pet-ownership-statistics.

Anderson, John D., Jr. "Ludwig Prandtl's Boundary Layer." *Physics Today* 58 (2005): 42–48.

Andersson, Malte. *Sexual Selection.* Princeton, NJ: Princeton University Press, 1994.

Anonymous. "Cause of the Black Spots on the Scales of Fish." In *Annual Record of Science and Industry for 1876,* edited by Spencer F. Baird et al., 332. New York: Harper and Brothers, 1877.

———. "Literary Notices: Science Sketches, by David Starr Jordan." *Popular Science Monthly* 32 (1888): 563.

———. "Scientific News." *English Mechanic and World of Science* 21 (1875): 10.

Aristotle. *History of Animals, in Ten Books.* Translated by R. Cresswell. London: H. G. Bohn, 1862.

Armistead, A. Wilson. "A Transfer of Leather Carp (*Cyprinus carpio*) from the Government Ponds at Washington, U.S.A., to Scotland." *Bulletin of the United States Fish Commission* 1 (1881): 341–42.

Audubon, John J. *Ornithological Biography.* Vol. 1. Philadelphia: E. L. Carey and A. Hart, 1832.

Avila, Vernon L. "A Field Study of Nesting Behavior of Male Bluegill Sunfish (*Lepomis macrochirus* Rafinesque)." *American Midland Naturalist* 96 (1976): 195–206.

Axelrod, Robert. *The Evolution of Cooperation.* New York: Basic Books, 1984.

Axelrod, Robert, and William D. Hamilton. "The Evolution of Cooperation." *Science* 211 (1981): 1390–96.

Baird, Spencer F. "Report of the Commissioner." In *United States Commission of Fish and Fisheries, Part II, Report of the Commissioner for 1872 and 1873,* i–xcii. Washington, DC: US Government Printing Office, 1874.

———. "Report of the Commissioner." In *United States Commission of Fish and Fisheries, Part III, Report of the Commissioner for 1873–4 and 1874–5,* vii–xlvi. Washington, DC: US Government Printing Office, 1876.

Baker, David H. "An Unusual Foreign Body: Catfish Spine." *Pediatric Radiology* 27 (1997): 585.

Baker, J. A., and S. A. Foster. "Observations on a Foraging Association between Two Freshwater Stream Fishes." *Ecology of Freshwater Fish* 3 (1994): 137–39.

Balon, Eugene K. "Origin and Domestication of the Wild Carp, *Cyprinus carpio*: From Roman Gourmets to the Swimming Flowers." *Aquaculture* 129 (1995): 3–48.

Barber, Iain, Stephen A. Arnott, Victoria A. Braithwaite, Jennifer Andrew, and Felicity A. Huntingford. "Indirect Fitness Consequences of Mate Choice in Sticklebacks: Offspring of Brighter Males Grow Slowly but Resist Parasitic Infections." *Proceedings of the Royal Society of London B: Biological Sciences* 268 (2001): 71–76.

Bart, Henry L., Jr., and Lawrence M. Page. "Morphology and Adaptive Significance of Fin Knobs in Egg-Clustering Darters." *Copeia* 1991 (1991): 80–86.

Beamish, F. William H., and Jo-Anne Jebbink. "Abundance of Lamprey Larvae and Physical Habitat." *Environmental Biology of Fishes* 39 (1994): 209–14.

Bean, Tarleton H. *The Fishes of Pennsylvania, with Description of the Species and Notes on Their Common Names, Distribution, Habits, Reproduction, Rate of Growth, and Mode of Capture.* Harrisburg, PA: Edwin K. Meyers, Binder, 1892.

Behrens, Roy R. "Revisiting Abbott Thayer: Non-scientific Reflections about Camouflage in Art, War and Zoology." *Philosophical Transactions of the Royal Society of London B: Biological Sciences* 364 (2009): 497–501.

Bennett, Keith D., and J. L. Fuller. "Determining the Age of the Mid-Holocene *Tsuga canadensis* (Hemlock) Decline, Eastern North America." *Holocene* 12 (2002): 421–29.

Bennett, Micah G., and Kevin W. Conway. "An Overview of North America's Diminutive Freshwater Fish Fauna." *Ichthyological Exploration of Freshwaters* 21 (2010): 63–72.

Berra, Tim M., and Ray-Jean Au. "Incidence of Black Spot Disease in Fishes in Cedar Fork Creek, Ohio." *Ohio Journal of Science* 78 (1978): 318–22.

Bhiry, Najat, and Louise Filion. "Mid-Holocene Hemlock Decline in Eastern North America Linked with Phytophagous Insect Activity." *Quaternary Research* 45 (1996): 312–20.

Birkhead, William S. "Toxicity of Stings of Ariid and Ictalurid Catfishes." *Copeia* 1972 (1972): 790–807.

Blake, Robert W. "Biomechanics of Rheotaxis in Six Teleost Genera." *Canadian Journal of Zoology* 84 (2006): 1173–86.

Boewe, C. "The Fall from Grace of That 'Base Wretch' Rafinesque." *Kentucky Review* 7 (1987): 39–53.

Bond, Gerard, William Showers, Maziet Cheseby, Rusty Lotti, Peter Almasi, Peter deMenocal, Paul Priore, Heidi Cullen, Irka Hajdas, and Georges Bonani. "A Pervasive Millennial-Scale Cycle in North Atlantic Holocene and Glacial Climates." *Science* 278 (1997): 1257–66.

Bonneau, Joseph L., and Dennis L. Scarnecchia. "The Zooplankton Community of a Turbid Great Plains (USA) Reservoir in Response to a Biomanipulation with Common Carp (*Cyprinus carpio*)." *Transactions of the Kansas Academy of Science* 117 (2014): 181–92.

Bonser, Richard H. C., and Peter P. Purslow. "The Young's Modulus of Feather Keratin." *Journal of Experimental Biology* 198 (1995): 1029–33.

Booth, Robert K., Simon Brewer, Maarten Blaauw, Thomas A. Minckley, and Stephen T. Jackson. "Decomposing the Mid-Holocene *Tsuga* Decline in Eastern North America." *Ecology* 93 (2012): 1841–52.

Boschung, Herbert T., and Richard L. Mayden. *Fishes of Alabama.* Washington, DC: Smithsonian Books, 2004.

Bosher, Bishop T., Scott H. Newton, and Michael L. Fine. "The Spines of the Channel Catfish, *Ictalurus punctatus,* as an Anti-predator Adaptation: An Experimental Study." *Ethology* 112 (2005): 188–95.

Bowdoin, Charles S. "Waterways of Innovation: The Marine Technological Advancements of America's Prohibition Era." MS thesis, East Carolina University, 2016.

Bowman, Milton L. "Life History of the Black Redhorse, *Moxostoma duquesnei* (Lesueur), in Missouri." *Transactions of the American Fisheries Society* 99 (1970): 546–59.

Brantley, Steven T., Chelcy Ford Miniat, Katherine J. Elliott, Stephanie H. Laseter, and James M. Vose. "Changes to Southern Appalachian Water Yield and Stormflow after Loss of a Foundation Species." *Ecohydrology* 8 (2015): 518–28.

Breder, Charles M., Jr. "The Reproductive Habits of the North American Sunfishes (Family Centrarchidae)." *Zoologica* 21 (1936): 1–48.

Brown, James M., and Naomi Weisstein. "A Spatial Frequency Effect on Perceived Depth." *Perception & Psychophysics* 44 (1988): 157–66.

Buck, Stanley E., III. "Insect Fauna Associated with Eastern Hemlock, *Tsuga canadensis* (L.), in the Great Smoky Mountains National Park." MS thesis, University of Tennessee, 2004.

Bumann, D., J. Krause, and D. Rubenstein. "Mortality Risk of Spatial Positions in Animal Groups: The Danger of Being in the Front." *Behaviour* 134 (1997): 1063–76.

Cahn, Alvin R. "The Effect of Carp on a Small Lake: The Carp as a Dominant." *Ecology* 10 (1929): 271–74.

Call, R. E. *The Life and Writings of Rafinesque.* Louisville, KY: John P. Morton, 1895.

Campbell, F. W., and L. Maffei. "Contrast and Spatial Frequency." *Scientific American* 231, no. 5 (1974): 106–15.

———. "The Influence of Spatial Frequency and Contrast on the Perception of Moving Patterns." *Vision Research* 21 (1981): 713–21.

Carey, Michael P., Beth L. Sanderson, Thomas A. Friesen, Katie A. Barnas, and Julian D. Olden. "Smallmouth Bass in the Pacific Northwest: A Threat to Native Species; A Benefit for Anglers." *Reviews in Fisheries Science* 19 (2011): 305–15.

Cargnelli, Luca M., and Bryan D. Neff. "Condition-Dependent Nesting in Bluegill Sunfish *Lepomis macrochirus.*" *Journal of Animal Ecology* 75 (2006): 627–33.

Carling, P. A. "The Nature of the Fluid Boundary Layer and the Selection of Parameters for Benthic Ecology." *Freshwater Biology* 28 (1992): 273–84.

Carlson, Rose L., and George V. Lauder. "Escaping the Flow: Boundary Layer Use by the Darter *Etheostoma tetrazonum* (Percidae) during Benthic Station Holding." *Journal of Experimental Biology* 214 (2011): 1181–93.

———. "Living on the Bottom: Kinematics of Benthic Station-Holding in Darter Fishes (Percidae: Etheostomatinae)." *Journal of Morphology* 271 (2010): 25–35.

Carlson, S. M., A. P. Hendry, and B. H. Letcher. "Growth Rate Differences between Resident Native Brook Trout and Non-native Brown Trout." *Journal of Fish Biology* 71 (2007): 1430–47.

Casselman, John M., and David K. Cairns, eds. *Eels at the Edge: Science, Status, and Conservation Concerns.* Proceedings of the 2003 International Eel Symposium. Bethesda, MD: American Fisheries Society, 2009.

Catovsky, S., N. M. Holbrook, and F. A. Bazzaz. "Coupling Whole-Tree Transpiration and Canopy Photosynthesis in Coniferous and Broad-Leaved Tree Species." *Canadian Journal of Forest Research* 32 (2002): 295–309.

Childress, Evan S., J. David Allan, and Peter B. McIntyre. "Nutrient Subsidies from Iteroparous Fish Migrations Can Enhance Stream Productivity." *Ecosystems* 17 (2014): 522–34.

Choi, Jin, Woo-Pyung Jeon, and Haecheon Choi. "Mechanism of Drag Reduction by Dimples on a Sphere." *Physics of Fluids* 18 (2006): 041702. https://doi.org/10.1063/1.2191848.

Church, Jarrod E., and Wayne C. Hodgson. "The Pharmacological Activity of Fish Venoms." *Toxicon* 40 (2002): 1083–93.

Cincotta, Daniel A., Douglas B. Chambers, and Terence Messinger. "Recent Changes in the Distribution of Fish Species in the New River Basin in West Virginia and Virginia." In *Proceedings of the New River Symposium, April 15–16, Boone, North Carolina*, 98–106. Glen Jean, WV: US National Park Service, 1999.

Cincotta, Daniel A., and Michael E. Hoeft. "Rediscovery of the Crystal Darter, *Ammocrypta asprella,* in the Ohio River Basin." *Brimleyana* 13 (1987): 133–36.

Clark, Nancy. "The Hermitage—Legacy of a Century-Old Inn." *Wonderful West Virginia* 43, no. 12 (1980): 2–6.

Clay, William M. *The Fishes of Kentucky.* Frankfort: Kentucky Department of Fish and Wildlife Resources, 1975.

Clinton, DeWitt. "Description of a New Species of Fish from the Hudson River (*Clupea hudsonia*)." *Annals of the Lyceum of Natural History of New York* 1 (1824): 49–50.

Cochran, Philip A. "Predation on Lampreys." *American Fisheries Society Symposium* 72 (2009): 139–51.

Cochran, Philip A., Devin D. Bloom, and Richard J. Wagner. "Alternative Reproductive Behaviors in Lampreys and Their Significance." *Journal of Freshwater Ecology* 23 (2008): 437–44.

Cochran, Philip A., and Mark A. Zoller. "'Willow Cats' for Sale? Madtoms (Genus *Noturus*) as Bait in the Upper Mississippi River Valley." *American Currents* 35, no. 2 (2009): 1–8.

Coffman, William P., Kenneth W. Cummins, and John C. Wuycheck. "Energy Flow in a Woodland Stream Ecosystem: I. Tissue Support Trophic Structure of the Autumnal Community." *Archiv für Hydrobiologie* 68 (1971): 232–76.

Cole, Leon J. "The German Carp in the United States." In *Appendix to the Report of the Commissioner of Fisheries to the Secretary of Commerce and Labor for the Year Ending June 30, 1904*, 523–641. Washington, DC: US Government Printing Office, 1905.

Colgan, Patrick W., William A. Nowell, Mart R. Gross, and James W. A. Grant. "Aggressive Habituation and Rim Circling in the Social Organization of Bluegill Sunfish (*Lepomis macrochirus*)." *Environmental Biology of Fishes* 4 (1979): 29–36.

Conservation Fisheries. "Diamond Darters Spawning at CFI 2010." Accessed July 10, 2019. https://www.youtube.com/watch?v=tQSvqXX_az8.

Cooke, Steven J., Christopher M. Bunt, Steven J. Hamilton, Cecil A. Jennings, Michael P. Pearson, Michale S. Cooperman, and Douglas F. Markle. "Threats, Conservation Strategies, and Prognosis for Suckers (Catostomidae) in North America: Insights from Regional Case Studies of a Diverse Family of Nongame Fishes." *Biological Conservation* 121 (2005): 317–31.

Coombs, Sheryl, Erik Anderson, Christopher B. Braun, and Mark Grosenbaugh. "The Hydrodynamic Footprint of a Benthic, Sedentary Fish in Unidirectional Flow." *Journal of the Acoustical Society of America* 122 (2007): 1227–37.

Coots, Carla, Paris Lambdin, Jerome Grant, and Rusty Rhea. "Diversity, Vertical Stratification and Co-occurrence Patterns of the Mycetophilid Community among Eastern Hemlock, *Tsuga canadensis* (L.) Carrière, in the Southern Appalachians." *Forests* 3 (2012): 986–96.

Cope, Edward D. "On the Distribution of Fresh-Water Fishes in the Allegheny Region of Southwestern Virginia." *Journal of the Academy of Natural Sciences of Philadelphia,* 2nd ser., 6, part 3, art. 5 (1869): 207–47.

———. "Synopsis of the Cyprinidae of Pennsylvania." *Transactions of the American Philosophical Society,* n.s., 13, part 3, art. 13 (1869): 351–99.

Cotchefer, Richard. "Report of the General Foreman of Hatcheries." In *Seventh Report of the Forest, Fish, and Game Commission of the State of New York,* 61. Albany, NY: J. B. Lyon, 1902.

Courtwright, Jennifer, and Christine L. May. "Importance of Terrestrial Subsidies for Native Brook Trout in Appalachian Intermittent Streams." *Freshwater Biology* 58 (2013): 2423–38.

Crivelli, Alain J. "The Destruction of Aquatic Vegetation by Carp: A Comparison between Southern France and the United States." *Hydrobiologia* 106 (1983): 37–41.

Cross, Frank B. *Handbook of Fishes of Kansas.* Miscellaneous Publications of the Museum of Natural History, University of Kansas 45. Lawrence, KS: Museum of Natural History, 1967.

Curry, R. Allen, and W. Scott MacNeill. "Population-Level Responses to Sediment during Early Life in Brook Trout." *Journal of the North American Benthological Society* 23 (2004): 140–50.

Daley, Michael J., Nathan G. Phillips, Cory Pettijohn, and Julian L. Hadley. "Water Use by Eastern Hemlock (*Tsuga canadensis*) and Black Birch (*Betula lenta*): Implications of Effects of the Hemlock Woolly Adelgid." *Canadian Journal of Forest Research* 37 (2007): 2031–40.

Dall, William H. *Spencer Fullerton Baird: A Biography, Including Selections from His Correspondence with Audubon, Agassiz, Dana, and Others.* London: J. B. Lippincott, 1915.

Daniels, Robert A. "Habitat of the Eastern Sand Darter, *Ammocrypta pellucida.*" *Journal of Freshwater Ecology* 8 (1993): 287–95.

———. "Significance of Burying in *Ammocrypta pellucida.*" *Copeia* 1989 (1989): 29–34.

Danisi, Thomas C. "Preserving the Legacy of Meriwether Lewis: The Letters of Samuel Latham Mitchill." *Quarterly Journal of the Lewis and Clark Foundation* 36 (2010): 8–11.

Darwin, Charles. *The Descent of Man, and Selection in Relation to Sex.* London: John Murray, 1871.

———. *On the Origin of Species by Means of Natural Selection, or the Preservation of Favoured Races in the Struggle for Life.* London: John Murray, 1859.

Davies, William E. *The Geology and Engineering Structures of the Chesapeake and Ohio Canal: An Engineering Geologist's Descriptions and Drawings.* Glen Echo, MD: C&O Canal Association, 1999.

Davis, H. B. "Gratifying Results of Propagating German Carp—Bream and Carp in Ponds Together—Table Qualities of Carp." *Bulletin of the United States Fish Commission* 2 (1882): 317–18.

Dawood, A. A., J. F. Price, and A. E. Reynolds Jr. "Utilization of Minced Sucker Flesh." *Journal of Food Quality* 6 (1983): 49–64.

Dean, Bashford. "The Early Development of Gar-Pike and Sturgeon." *Journal of Morphology* 11 (1895): 1–62.

Dean, Bashford, and Francis B. Sumner. "Notes on the Spawning Habitats of the Brook Lamprey (*Petromyzon wilderi*)." *Transactions of the New York Academy of Sciences* 16 (1897): 321–24.

Dean, Brian, and Bharat Bhushan. "Shark-Skin Surfaces for Fluid-Drag Reduction in Turbulent Flow: A Review." *Philosophical Transactions of the Royal Society of London A* 368 (2010): 4775–806.

Deane, Ruthven. "Unpublished Letters of John James Audubon and Spencer F. Baird." *Auk* 23 (1906): 318–34.

———. "Unpublished Letters of John James Audubon and Spencer F. Baird." *Auk* 24 (1907): 53–70.

DeKay, James E. *Zoology of New-York, or the New-York Fauna; Part IV, Fishes.* Albany, NY: W. and A. White and J. Visscher, 1842.

Delcourt, Johann, and Pascal Poncin. "Shoals and Schools: Back to the Heuristic Definitions and Quantitative References." *Reviews in Fish Biology and Fisheries* 22 (2012): 595–619.

Denton, George H., and Wibjörn Karlén. "Holocene Climatic Variations—Their Pattern and Possible Cause." *Quaternary Research* 3 (1973): 155–205.

DeWald, Lynn, and Margaret A. Wilzbach. "Interactions between Native Brook Trout and Hatchery Brown Trout—Effects on Habitat Use, Feeding, and Growth." *Transactions of the American Fisheries Society* 121 (1992): 287–96.

DeWalle, David R., and Bryan R. Swistock. "Causes of Episodic Acidification in Five Pennsylvania Streams on the Northern Appalachian Plateau." *Water Resources Research* 30 (1994): 1955–63.

Dilling, Carla, Paris Lambdin, Jerome Grant, and Lee Buck. "Insect Guild Structure Associated with Eastern Hemlock in the Southern Appalachians." *Environmental Entomology* 36 (2007): 1408–14.

Dixon, Douglas A., ed. *Biology, Management, and Protection of Catadromous Eels.* Symposium 33. Bethesda, MD: American Fisheries Society, 2003.

Docker, Margaret F. "A Review of the Evolution of Nonparasitism in Lampreys and an Update of the Paired Species Concept." In *Biology, Management, and Conservation of Lampreys in North America,* edited by Larry R. Brown, Shawn D. Chase, Matthew G. Mesa, Richard J. Beamish, and Peter B. Moyle, 71–114. Bethesda, MD: American Fisheries Society, 2009.

Docker, Margaret F., and F. William H. Beamish. "Growth, Fecundity, and Egg Size of Least Brook Lamprey, *Lampetra aepyptera*." *Environmental Biology of Fishes* 31 (1991): 219–27.

Docker, Margaret F., Nicholas E. Mandrak, and Daniel D. Heath. "Contemporary Gene Flow between 'Paired' Silver (*Ichthyomyzon unicuspis*) and Northern Brook (*I. fossor*) Lampreys: Implications for Conservation." *Conservation Genetics* 13 (2012): 823–35.

Doctor, Daniel H., David J. Weary, David K. Brezinski, Randall C. Orndorff, and Lawrence E. Spangler. "Karst of the Mid-Atlantic Region in Maryland, West Virginia, and Virginia." In *Tripping from the Fall Line: Field Excursions for the GSA Annual Meeting, Baltimore, 2015, Field Guide 40,* edited by David K. Brezinski, Jeffrey P. Halka, and Richard A. Ortt Jr., 425–84. [Boulder, CO]: Geological Society of America, 2015.

Dodd, L. E., Z. Cornett, A. Smith, and L. K. Rieske. "Variation in Lepidopteran Occurrence in Hemlock-Dominated and Deciduous-Dominated Forests of Central Appalachia." *Great Lakes Entomologist* 46 (2003): 1–12.

Dominey, Wallace J. "Female Mimicry in Bluegill Sunfish—a Genetic Polymorphism?" *Nature* 284 (1980): 546–48.

Dorgan, Kelly M., Peter A. Jumars, Bruce D. Johnson, and Bernard P. Boudreau. "Macrofaunal Burrowing: The Medium Is the Message." *Oceanography and Marine Biology: An Annual Review* 44 (2006): 85–121.

Douglas, Neil H. "A New Record Size for Darters." *Proceedings of the Louisiana Academy of Sciences* 31 (1968): 41–42.

Dugatkin, Lee A. "Do Guppies Play TIT FOR TAT during Predator Inspection Visits?" *Behavioral Ecology and Sociobiology* 25 (1988): 395–99.

———. "Tendency to Inspect Predators Predicts Mortality Risk in the Guppy (*Poecilia reticulata*)." *Behavioral Ecology* 3 (1992): 124–27.

Dugatkin, Lee A., and Jean-Guy J. Godin. "Predator Inspection, Shoaling and Foraging under Predation Hazard in the Trinidadian Guppy, *Poecilia reticulata.*" *Environmental Biology of Fishes* 34 (1992): 265–76.

———. "Prey Approaching Predators: A Cost-Benefit Perspective." *Annales Zoologici Fennici* 29 (1992): 233–52.

Duigon, Mark T. *Phase 2 Study of the Area Contributing Groundwater to the Spring Supplying the A. M. Powell State Fish Hatchery, Washington County, Maryland.* Maryland Geological Survey, Open-File Report No. 2008-02-18. [Baltimore]: Maryland Geological Survey, 2008.

Eaton, John G., and Robert M. Scheller. "Effects of Climate Warming on Fish Thermal Habitat in Streams of the United States." *Limnology and Oceanography* 41 (1996): 1109–15.

Eckhart, Leopold, Luisa Dalla Valle, Karin Jaeger, Claudia Ballaun, Sandra Szabo, Alessia Nardi, Maria Buchberger, Marcela Hermann, Lorenzo Alibardi, and Erwin Tschachler. "Identification of Reptilian Genes Encoding Hair Keratin–Like Proteins Suggests a New Scenario for the Evolutionary Origin of Hair." *Proceedings of the National Academy of Sciences* 105 (2008): 18419–23.

Edwards, Linden F. "The Protactile Apparatus of the Mouth of the Catostomid Fishes." *Anatomical Record* 33 (1926): 257–70.

Egge, Jacob J. D., and Andrew M. Simons. "Evolution of Venom Delivery Structures in Madtom Catfishes (Siluriformes: Ictaluridae)." *Biological Journal of the Linnean Society* 102 (2011): 115–29.

Ellison, Aaron M., Michael S. Bank, Barton D. Clinton, Elizabeth A. Colburn, Katherine Elliott, Chelcy R. Ford, David R. Foster et al. "Loss of Foundation Species: Consequences for the Structure and Dynamics of Forested Ecosystems." *Frontiers in Ecology and the Environment* 3 (2005): 479–86.

Emmett, Bridget, and Philip A. Cochran. "The Response of a Piscivore (*Micropterus salmoides*) to a Venomous Prey Species (*Noturus gyrinus*)." *Journal of Freshwater Ecology* 25 (2010): 475–79.

Endler, John A. *Natural Selection in the Wild.* Princeton, NJ: Princeton University Press, 1986.

Ensign, Scott H., and Martin W. Doyle. "Nutrient Spiraling in Streams and River Networks." *Journal of Geophysical Research* 111 (2006): G04009, https://doi.org/10.1029/2005JG000114.

Eoff, John. "On the Habits of the Black Bass of the Ohio (*Grystes fasciatus*)." In *Ninth Annual Report of the Board of Regents of the Smithsonian Institution,* 289–90. Washington, DC: Beverley Tucker, Senate Printer, 1855.

Eschner, Kat. "This Obscure Fishing Book Is One of the Most Reprinted English Books Ever." *Smithsonian Magazine,* August 9, 2017. www.smithsonianmag.com.

Evans, Daniel M., W. Michael Aust, C. Andrew Dolloff, Ben S. Templeton, and John A. Peterson. "Eastern Hemlock Decline in Riparian Areas from Maine to Alabama." *Northern Journal of Applied Forestry* 28 (2011): 97–104.

Evans, Daniel M., C. Andrew Dolloff, W. Michael Aust, and Amy M. Villamagna. "Effects of Eastern Hemlock Decline on Large Wood Loads in Streams of the Appalachian Mountains." *Journal of the American Water Resources Association* 48 (2012): 266–76.

Evans, Justin D., and Lawrence M. Page. "Distribution and Relative Size of the Swim Bladder in *Percina,* with Comparisons to *Etheostoma, Crystallaria,* and *Ammocrypta* (Teleostei: Percidae)." *Environmental Biology of Fishes* 66 (2003): 61–65.

Eyler, Sheila M. "Timing and Survival of American Eels Migrating Past Hydroelectric Dams on the Shenandoah River." PhD diss., West Virginia University, 2014.

Eyler, Sheila M., Stuart A. Welsh, David R. Smith, and Mary M. Rockey. "Downstream Passage and Impact of Turbine Shutdowns on Survival of Silver American Eels at Five Hydroelectric Dams on the Shenandoah River." *Transactions of the American Fisheries Society* 145 (2016): 964–76.

Facey, D. E., and G. D. Grossman. "The Metabolic Cost of Maintaining Position for Four North American Stream Fishes: Effects of Season and Velocity." *Physiological Zoology* 63 (1990): 757–76.

Facey, Douglas E., and Michael J. Van Den Avyle. *Species Profiles: Life Histories and Environmental Requirements of Coastal Fishes and Invertebrates (North Atlantic)–American Eel.* Washington, DC: US Fish and Wildlife Service, 1987.

Fiedel, S. "Older Than We Thought: Implications of Corrected Dates for Paleoindians." *American Antiquity* 64 (1999): 95–115.

Fine, Michael L., John P. Friel, David McElroy, Charles B. King, Kathryn E. Loesser, and Scott Newton. "Pectoral Spine Locking and Sound Production in the Channel Catfish *Ictalurus punctatus.*" *Copeia* 1997 (1997): 777–90.

Fisher, James. "Evolution and Bird Sociality." In *Evolution as a Process,* edited by J. Huxley, A. C. Hardy, and E. B. Ford, 71–83. London: Allen and Unwin, 1954.

Fisher, Stuart G., and Gene E. Likens. "Energy Flow in Bear Brook, New Hampshire: An Integrative Approach to Stream Ecosystem Metabolism." *Ecological Monographs* 43 (1973): 421–39.

Fitzpatrick, T. J. *Rafinesque: A Sketch of His Life with Bibliography.* Des Moines: Historical Department of Iowa, 1911.

Forbes, L. Scott. "Prey Defences and Predator Handling Behaviour: The Dangerous Prey Hypothesis." *Oikos* 55 (1989): 155–58.

Forbes, Stephen A., and Robert E. Richardson. *The Fishes of Illinois.* 2nd ed. Springfield: Illinois State Journal, State Printers, 1920.

Ford, Chelcy R., and James M. Vose. "*Tsuga canadensis* (L.) Carr: Mortality Will Impact Hydrological Processes in Southern Appalachian Forest Ecosystems." *Ecological Applications* 17 (2007): 1156–67.

Fowler, James F., and Charles A. Taber. "Food Habits and Feeding Periodicity in Two Sympatric Stonerollers (Cyprinidae)." *American Midland Naturalist* 113 (1985): 217–24.

Francis, John W. *Reminiscences of Samuel Latham Mitchill, M.D., LL.D.* New York: John F. Trow, 1859.

Frith, H. Russ, and Robert W. Blake. "The Mechanical Power Output and Hydromechanical Efficiency of Northern Pike (*Esox lucius*) Fast-Starts." *Journal of Experimental Biology* 198 (1995): 1863–73.

Gajewski, Konrad, and Andre E. Viau. "Abrupt Climate Changes during the Holocene across North America from Pollen and Paleolimnological Records." In *Abrupt Climate Change: Mechanisms, Patterns, and Impacts,* edited by Harunur Rashid, Leonid Polyak, and Ellen Mosley-Thompson, 173–83. Geophysical Monograph Series 193. Washington, DC: American Geophysical Union, 2011.

Gasper, Donald C. "West Virginia Trout Streams: Target for Acid Precipitation." In *Confluence 1983,* edited by Rick Webb, 5–11. Morgantown, WV: West Virginia Mountain Stream Monitors, 1983.

Geist, Valerius. "The Evolution of Horn-Like Organs." *Behavior* 27 (1966): 175–214.

Gende, Scott M., Richard T. Edwards, Mary F. Willson, and Mark S. Wipfli. "Pacific Salmon in Aquatic and Terrestrial Ecosystems: Pacific Salmon Subsidize Freshwater and Terrestrial Ecosystems through Several Pathways, Which Generates Unique Management and Conservation Issues but Also Provides Valuable Research Opportunities." *BioScience* 52 (2002): 917–98.

George, Kelly A., Kristina M. Slagle, Robyn S. Wilson, Steven J. Moeller, and Jeremy T. Bruskotter. "Changes in Attitudes toward Animals in the United States from 1978." *Biological Conservation* 201 (2016): 237–42.

German, Donovan P. "Do Herbivorous Minnows Have 'Plug-Flow Reactor' Guts? Evidence from Digestive Enzyme Activities, Gastrointestinal Fermentation, and Luminal Nutrient Concentrations." *Journal of Comparative Physiology B* 179 (2009): 759–71.

Gibson, Isaac. "Conservation Concerns for the Candy Darter (*Etheostoma osburni*) with Implications Related to Hybridization." MS thesis, West Virginia University, 2017.

Gibson, Isaac, Amy B. Welsh, Stuart A. Welsh, and Daniel A. Cincotta. "Genetic Swamping and Possible Species Collapse: Tracking Introgression between the Native Candy Darter and Introduced Variegate Darter." *Conservation Genetics* 20 (2019): 287–98.

Godin, Jean-Guy J., and Shelley L. Crossman. "Hunger-Dependent Predator Inspection and Foraging Behaviours in the Threespine Stickleback (*Gasterosteus aculeatus*) under Predation Risk." *Behavioral Ecology and Sociobiology* 34 (1994): 359–66.

Goode, G. Brown. *A Review of the Fishery Industries of the United States and the Work of the U.S. Fish Commission.* London: William Clowes and Sons, 1883.

Goodwin, Kevin R., and Paul L. Angermeier. "Demographic Characteristics of American Eel in the Potomac River Drainage, Virginia." *Transactions of the American Fisheries Society* 132 (2003): 524–35.

Gottschalk, John S. "The Introduction of Exotic Animals into the United States." In *Towards a New Relationship of Man and Nature in Temperate Lands,* 124–40. Morges, Switzerland: International Union for Conservation of Nature and Natural Resources, 1967.

Gould, Stephen J. "Red Wings in the Sunset." *Natural History* 94 (1985): 12–24.

Gowanloch, James N. "Gars, Killers of Game and Food Fish." *Louisiana Conservation Review* 8 (1939): 44–46, 50.

Greeley, John R. "The Spawning Habits of Brook, Brown, and Rainbow Trout and the Problem of Egg Predation." *Transactions of the American Fisheries Society* 62 (1932): 239–48.

Gregory, William K. "Memorial of Bashford Dean, 1867–1928." In *The Bashford Dean Memorial Volume: Archaic Fishes, Part 1,* edited by E. W. Gudger, 1–42. New York: American Museum of Natural History, 1930.

Grice, Hannah. "Effects on *Campostoma oligolepis* Digestive Morphology and Gut Microbiota Composition across a Gradient of Urbanization." MS thesis, Kennesaw State University, 2017.

Gross, Mart R. "Cuckoldry in Sunfishes (*Lepomis:* Centrarchidae)." *Canadian Journal of Zoology* 57 (1979): 1507–9.

———. "Evolution of Alternative Reproductive Strategies: Frequency-Dependent Sexual Selection in Male Bluegill Sunfish." *Transactions of the Royal Society of London B* 332 (1991): 59–66.

———. "Sneakers, Satellites and Parentals: Polymorphic Mating Strategies in North-American Sunfishes." *Zeitschrift für Tierpsychologie* 60 (1982): 1–26.

Gross, Mart R., and Eric L. Charnov. "Alternative Male Life Histories in Bluegill Sunfish." *Proceedings of the National Academy of Sciences of the United States of America* 77 (1980): 6937–40.

Gross, Mart R., and Joe Repka. "Stability with Inheritance in the Conditional Strategy." *Journal of Theoretical Biology* 192 (1998): 445–53.

Haas, Jean Nicolas, and John H. McAndrews. "The Summer Drought Related Hemlock (*Tsuga canadensis*) Decline in Eastern North America 5700 to 5100 Years Ago." In *Proceedings: Sustainable Management of Hemlock Ecosystems in Eastern North America,* edited by Katherine A. McManus, Kathleen S. Shields, and Dennis R. Souto, 81–88. Durham, NH: US Department of Agriculture, 2000.

Hack, John T. *Geomorphology of the Shenandoah Valley Virginia and West Virginia and Origin of the Residual Ore Deposits.* US Geological Survey Professional Paper 484. Washington, DC: US Government Printing Office, 1965.

Hadley, Julian L. "Understory Microclimate and Photosynthetic Response of Saplings in an Old-Growth Eastern Hemlock (*Tsuga canadensis* L.) Forest." *Ecoscience* 7 (2000): 66–72.

Halstead, Bruce W., Leonard S. Kuninobu, and Harold G. Hebard. "Catfish Stings and the Venom Apparatus of the Mexican Catfish, *Galeichthys felis* (Linnaeus)." *Transactions of the American Microscopical Society* 72 (1953): 297–314.

Hamilton, William D. "Geometry for the Selfish Herd." *Journal of Theoretical Biology* 31 (1971): 295–311.

Hamilton, William D., and Marlene Zuk. "Heritable True Fitness and Bright Birds: A Role for Parasites?" *Science* 218 (1982): 384–87.

Hammond, Stephen D., and Stuart A. Welsh. "Seasonal Movements of Large Yellow American Eels Downstream of a Hydroelectric Dam, Shenandoah River, West Virginia." In *Eels at the Edge: Science, Status, and Conservation Concerns,* edited by John M. Casselman and David K. Cairns, 309–23. Proceedings of the 2003 International Eel Symposium. Bethesda, MD: American Fisheries Society, 2009.

Hankin, James, and David B. Wake. "Miniaturization of Body Size: Organismal Consequences and Evolutionary Significance." *Annual Review of Ecology and Systematics* 24 (1993): 501–19.

Hardisty, Martin W. "The Growth of Larval Lampreys." *Journal of Animal Ecology* 30 (1961): 357–71.

Hardisty, Martin W., and Ian C. Potter. *The Biology of Lampreys.* Vol. 1. London: Academic, 1971.

Hargrave, Chad W., Raul Ramirez, Melody Brooks, Michael A. Eggleton, Katherine Sutherland, Raelynn Deaton, and Heather Galbraith. "Indirect Food Web Interactions Increase Growth of an Algivorous Stream Fish." *Freshwater Biology* 51 (2006): 1901–10.

Haro, Alex, William Richkus, Kevin Whalen, Alex Hoar, W.-Dieter Busch, Sandra Lary, Tim Brush, and Douglas Dixon. "Population Decline of the American Eel: Implications for Research and Management." *Fisheries* 25 (2000): 7–16.

Harper, David G., and Robert W. Blake. "Prey Capture and Fast-Start Performance of Northern Pike *Esox lucius*." *Journal of Experimental Biology* 155 (1991): 175–92.

Hayes, Daniel B., William W. Taylor, and James C. Schneider. "Response of Yellow Perch and the Benthic Invertebrate Community to a Reduction in the Abundance of White Suckers." *Transactions of the American Fisheries Society* 121 (1992): 36–53.

Helfman, Gene S., and Jennifer B. Clark. "Rotational Feeding: Overcoming Gape-Limited Foraging in Anguillid Eels." *Copeia* 1986 (1986): 679–85.

Hendrick, Robert. "A Raven in a Coal Scuttle: Theodore Roosevelt and the Animal Coloration Controversy." *American Biology Teacher* 57 (1995): 14–20.

Henshall, James A. *Book of the Black Bass Comprising Its Complete Scientific and Life History Together with a Practical Treatise on Angling and Fly Fishing and a Full Description of Tools, Tackle and Implements.* Cincinnati: Robert Clarke, 1881.

Herlihy, Alan T., Philip R. Kaufmann, and Mark E. Mitch. "Regional Estimates of Acid Mine Drainage Impact on Streams in the Mid-Atlantic and Southeastern United States." *Water, Air, and Soil Pollution* 50 (1990): 91–107.

Herrick, Francis H. *Audubon the Naturalist: A History of His Life and Time.* Vol. 1. New York: D. Appleton, 1917.

Hessel, Rudolph. "The Carp, and Its Culture in Rivers and Lakes: And Its Introduction into America." In *United States Commission of Fish and Fisheries, Part IV, Report of the Commissioner for 1875–1876,* 865–900. Washington, DC: US Government Printing Office, 1878.

Hessl, Amy, and Neil Pederson. "Hemlock Legacy Project (HeLP): A Paleoecological Requiem for Eastern Hemlock." *Progress in Physical Geography* 37 (2012): 114–29.

Hobba, William A., Jr., Donald W. Fisher, F. J. Pearson Jr., and Joseph C. Chemerys. *Hydrology and Geochemistry of Thermal Springs of the Appalachians.* US Geological Survey Professional Paper 1044-E. Washington, DC: US Government Printing Office, 1979.

Hofbauer, Josef, and Karl Sigmund. *Evolutionary Games and Populations Dynamics.* Cambridge: Cambridge University Press, 1988.

Hoffman, Glen L. *Parasites of North American Freshwater Fishes.* 2nd ed. Ithaca, NY: Cornell University Press, 1999.

———. *Synopsis of Strigeoidea (Trematoda) of Fishes and Their Life Cycles.* Fishery Bulletin 175. Washington, DC: US Government Printing Office, 1960.

Hogan, Benedict G., Innes C. Cuthill, and Nicholas E. Scott-Samuel. "Dazzle Camouflage and the Confusion Effect: The Influence of Varying Speed on Target Tracking." *Animal Behaviour* 123 (2017): 349–53.

Holey, Mark, Bruce Hollender, Mark Imhof, Roman Jesien, Richard Konopacky, Michael Toneys, and Daniel Coble. "Never Give a Sucker an Even Break." *Fisheries* 4, no. 1 (1979): 2–6.

Holloway, Ancil D. "Notes on the Life History and Management of the Shortnose and Longnose Gars in Florida Waters." *Journal of Wildlife Management* 18 (1954): 438–49.

Hoogland, R., Desmond Morris, and Nikolaas Tinbergen. "The Spines of Stick-lebacks (*Gasterosteus* and *Pygosteus*) as Means of Defence against Predators (*Perca* and *Esox*)." *Behaviour* 10 (1956): 205–36.

Hoquet, Thierry, and Michael Levandowsky. "Utility vs Beauty: Darwin, Wallace and the Subsequent History of the Debate on Sexual Selection." In *Current Perspectives on Sexual Selection*, edited by Thierry Hoquet, 19–44. Dordrecht, The Netherlands: Springer, 2015.

Hou, Juzhi, Yongsong Huang, Bryan N. Shuman, W. Wyatt Oswald, and David R. Foster. "Abrupt Cooling Repeatedly Punctuated Early-Holocene Climate in Eastern North America." *Holocene* 22 (2012): 525–29.

How, Martin J., and Johannes M. Zanker. "Motion Camouflage Induced by Zebra Stripes." *Zoology* 117 (2014): 163–70.

Hubbs, Carl L. "History of Ichthyology in the United States after 1850." *Copeia* 1964 (1964): 42–60.

Hubbs, Carl L., and Claude W. Hibbard. "*Ictalurus lambda*, a New Catfish, Based on a Pectoral Spine from the Lower Pliocene of Kansas." *Copeia* 1951 (1951): 8–14.

Hubbs, Carl L., and Milton B. Trautman. "*Poecilichthys osburni*, a New Darter from the Upper Kanawha River System in Virginia and West Virginia." *Ohio Journal of Science* 32 (1932): 31–38.

Hudy, Mark, Teresa Thieling, Nathaniel Gillespie, and Eric P. Smith. "Distribution, Status, and Land Use Characteristics of Subwatersheds within the Native Range of Brook Trout in the Eastern United States." *North American Journal of Fisheries Management* 28 (2008): 1069–85.

Hughes, Anna E., Jolyon Troscianko, and Martin Stevens. "Motion Dazzle and the Effects of Target Patterning on Capture Success." *BMC Evolutionary Biology* 14 (2014): 201.

Hughes, Ted. *Lupercal.* London: Faber and Faber, 1960.

Hunter, John R. "The Reproductive Behavior of the Green Sunfish *Lepomis cyanellus*." *Zoologica* 48 (1963): 13–24.

Hussakof, L. 1914. "Fishes Swallowed by Gar Pike." *Copeia* 1914, no. 11 (1914): 2.

Iguchi, Kei'ichiro, Taiga Yodo, and Naoto Matsubara. "Spawning and Brood Defense of Smallmouth Bass under the Process of Invasion into a Novel Habitat." *Environmental Biology of Fishes* 70 (2004): 219–25.

Ingram, William, and Eugene P. Odum. "Nests and Behavior of *Lepomis gibbosus* (Linne) in Lincoln Pond, Rensselaerville, N.Y." *American Midland Naturalist* 226 (1941): 182–93.

Jenkins, Robert E., and Noel M. Burkhead. *Freshwater Fishes of Virginia.* Bethesda, MD: American Fisheries Society, 1994.

Jenkins, Robert E., and D. J. Jenkins. "Reproductive Behavior of the Greater Red-horse, *Moxostoma valenciennesi*, in the Thousand Islands Region." *Canadian Field-Naturalist* 94 (1980): 426–30.

Jessop, Brian M. "Geographic Effects on American Eel (*Anguilla rostrata*) Life History Characteristics and Strategies." *Canadian Journal of Fisheries Aquatic Sciences* 67 (2010): 326–46.

Johnson, Fritz H. "Responses of Walleye (*Stizostedion vitreum vitreum*) and Yellow Perch (*Perca flavescens*) Populations to Removal of White Sucker (*Catostomus commersoni*) from a Minnesota Lake, 1966." *Journal of the Fisheries Board of Canada* 34 (1977): 1633–42.

Johnson, James H., and David S. Dropkin. "Piscivory by the Central Stoneroller *Campostoma anomalum*." *Journal of the Pennsylvania Academy of Science* 66, no. 2 (1992): 90–91.

Johnston, Carol E. "The Benefit to Some Minnows of Spawning in the Nests of Other Species." *Environmental Biology of Fishes* 40 (1994): 213–18.

———. "Nest Association in Fishes: Evidence for Mutualism." *Behavioral Ecology and Sociobiology* 35 (1994): 379–83.

———. "Spawning Activities of *Notropis chlorocephalus, Notropis chiliticus,* and *Hybopsis hypsinotus,* Nest Associates of *Nocomis leptocephalus* in the Southeastern United States, with Comments on Nest Association (Cypriniformes, Cyprinidae)." *Brimleyana* 17 (1991): 77–88.

Johnston, William D. *History of the Library of Congress.* Vol. 1, *1800–1864.* Washington, DC: US Government Printing Office, 1904.

Jones, Clive G., John H. Lawton, and Moshe Shachak. "Positive and Negative Effects of Organisms as Physical Ecosystem Engineers." *Ecology* 78 (1997): 1946–57.

Jordan, David S. "A Catalogue of the Fishes of Illinois." *Bulletin of the Illinois State Laboratory of Natural History* 1 (1878): 38–70.

———. "Concerning the Fishes of the Ichthyologia Ohiensis." *Bulletin of the Buffalo Society of Natural Sciences* 3 (1877): 91–97.

———. *The Days of a Man: Being Memories of a Naturalist, Teacher, and Minor Prophet of Democracy.* Vol. 1. Yonkers-on-Hudson, NY: World Book, 1922.

———. "Rafinesque." *Popular Science Monthly* 29 (1886): 212–21.

———. *Review of Rafinesque's Memoirs on North American Fishes.* Bulletin of the United States National Museum, no. 9. Washington, DC: US Government Printing Office, 1877.

———. *Science Sketches.* Chicago: A. C. McClurg, 1887.

———. "Spencer Fullerton Baird and the United States Fish Commission." *Scientific Monthly* 17 (1923): 97–107.

Jordan, David S., and Herbert E. Copeland. "The Sand Darter." *American Naturalist* 11 (1877): 86–88.

Jordan, David S., and Barton W. Evermann. "Description of a New Species of Fish from Tippecanoe River, Indiana." *Proceedings of the United States National Museum* 13 (1890): 3–4.

Jordan, David S., and Charles H. Gilbert. *Synopsis of the Fishes of North America.* Bulletin of the United States National Museum, no. 16. Washington, DC: US Government Printing Office, 1882.

Kaemingk, Mark A., Jeffrey C. Jolley, Craig P. Paukert, David W. Willis, Kjetil Henderson, Richard S. Holland, Greg A. Wanner, and Mark L. Lindvall. "Common Carp Disrupt Ecosystem Structure and Function through Middle-Out

Effects." *Marine and Freshwater Research* 68, no. 4 (2016): 718–31. https://doi
.org/10.1071/MF15068.

Kemp, Emory L. *The Great Kanawha Navigation*. Pittsburgh: University of Pittsburgh Press, 2000.

Kendall, William C. "Notes on *Percopsis guttatus* Agassiz and *Salmo omiscomaycus* Walbaum." *Proceedings of the Biological Society of Washington* 24 (1911):
45–51.

Kidd, Aline H., and Robert M. Kidd. "Benefits, Problems, and Characteristics of
Home Aquarium Owners." *Psychological Reports* 84 (1999): 998–1004.

Kingsland, Sharon. "Abbott Thayer and the Protective Coloration Debate." *Journal
of the History of Biology* 11 (1978): 223–44.

Kirkwood, James J. *Waterway to the West*. Washington, DC: Eastern National Park
and Monument Association, 1963.

Kirtland, Jared P. "Descriptions of Four New Species of Fishes." *Boston Journal of
Natural History* 3 (1840): 273–77.

Kleindl, Paige M., Fred D. Tucker, Michael G. Commons, Robert G. Verb, and Leslie A. Riley. "Influences of a *Tsuga canadensis* (L.) Carriere (Eastern Hemlock)
Riparian Habitat on a Lotic Benthic Community." *Northeastern Naturalist* 23
(2016): 555–70.

Knapp, Roland A., and Robert Craig Sargent. "Egg-Mimicry as a Mating Strategy
in the Fantail Darter: Females Prefer Males with Eggs." *Behavioral Ecology and
Sociobiology* 25 (1989): 321–26.

Knouft, Jason H., Lawrence M. Page, and Michael J. Plewa. "Antimicrobial Egg
Cleaning by the Fringed Darter (Perciformes: Percidae: *Etheostoma cros-
sopterum*): Implications of a Novel Component of Parental Care in Fishes."
Proceedings of the Royal Society of London B 270 (2003): 2405–11.

Kohlstedt, Sally G. "Nature, Not Books: Scientists and the Origins of the Nature-
Study Movement in the 1890s." *Isis* 96 (2005): 324–52.

Kozar, Mark D., and David J. Weary. *Hydrogeology and Ground-Water Flow in the
Opequon Creek Watershed Area, Virginia and West Virginia*. US Geological
Survey Scientific Investigations Report 2009-5153. Reston, VA: US Geological
Survey, 2009.

Kraak, Sarah B. M., and Ton G. G. Groothuis. "Female Preference for Nests with Eggs
Is Based on the Presence of the Eggs Themselves." *Behaviour* 131 (1994): 189–206.

Kraatz, Walter C. "The Intestine of the Minnow *Campostoma anomalum* (Rafinesque), with Special Reference to the Development of Its Coiling." *Ohio Journal of Science* 24 (1924): 265–98.

Krapfl, Kurt J., Eric J. Holzmueller, and Michael A. Jenkins. "Early Impacts of
Hemlock Woolly Adelgid in *Tsuga canadensis* Forest Communities of the
Southern Appalachian Mountains." *Journal of the Torrey Botanical Society* 138
(2011): 93–106.

Krause, Jens, and Jean-Guy J. Godin. "Influence of Parasitism on Shoal Choice
in the Banded Killifish (*Fundulus diaphanus*, Teleostei, Cyprinodontidae)."
Ethology 102 (1996): 40–49.

Krause, Jens, Jean-Guy J. Godin, and Dan Rubenstein. "Group Choice as a Function of Group Size Differences and Assessment Time in Fish: The Influence of Species Vulnerability to Predation." *Ethology* 104 (1998): 68–74.

Lacépède, Bernard G. *Histoire naturelle des poissons.* Vol. 4. Paris: Chez Plassan, 1802.

———. *Histoire naturelle des poissons.* Vol. 5. Paris: Plassan, 1803.

Lachner, Ernest A. "Studies of the Biology of the Cyprinid Fishes of the Chub Genus *Nocomis* of Northeastern United States." *American Midland Naturalist* 48 (1952): 433–66.

Lachner, Ernest A., and Robert E. Jenkins. "Systematics, Distribution, and Evolution of the Chub Genus *Nocomis* (Cyprinidae) in the Southwestern Ohio River Basin, with the Description of a New Species." *Copeia* 1967 (1967): 557–80.

———. "Systematics, Distribution, and Evolution of the Chub Genus *Nocomis* Girard (Pisces, Cyprinidae) of the Eastern United States, with Descriptions of New Species." *Smithsonian Contributions to Zoology* 85 (1971): 1–97.

———. "Systematics, Distribution, and Evolution of the *Nocomis biguttatus* Species Group (Family Cyprinidae: Pisces) with a Description of a New Species from the Ozark Upland." *Smithsonian Contributions to Zoology* 91 (1971): 1–28.

Lagler, Karl F., and Frances V. Hubbs. "Food of the Long-Nosed Gar (*Lepisosteus osseus oxyurus*) and the Bowfin (*Amia calva*) in Southern Michigan." *Copeia* 1940 (1940): 239–41.

Lagler, Karl F., Carl B. Obrecht, and George V. Harry. "The Food and Habits of Gars (*Lepisosteus* spp.) Considered in Relationship to Fish Management." *Investigations of Indiana Lakes and Streams* 2 (1942): 117–35.

Lagrée, Pierre-Yves. "Boundary Layer Separation and Asymptotics from 1904 to 1969." *Comptes Rendus Mecanique* 345 (2017): 613–19.

Lake, Charles T. "The Life History of the Fan-Tailed Darter *Catonotus flabellaris flabellaris* (Rafinesque)." *American Midland Naturalist* 17 (1936): 816–30.

Lauder, George V., Dylan K. Wainwright, August G. Domel, James C. Weaver, Li Wen, and Katia Bertoldi. "Structure, Biomimetics, and Fluid Dynamics of Fish Skin Surfaces." *Physical Review Fluids* 1 (2016): 060502. https://doi.org/10.1103/PhysRevFluids.1.060502.

Lennon, Robert E., and Phillip S. Parker. "The Stoneroller, *Campostoma anomalum* (Rafinesque), in Great Smoky Mountains National Park." *Transactions of the American Fisheries Society* 89 (1960): 263–70.

Leopold, Aldo. *A Sand County Almanac: And Sketches Here and There.* New York: Oxford University Press, 1949.

Lewis, Jennifer S., and William W. Schroeder. "Mud Plume Feeding, a Unique Foraging Behavior of the Bottlenose Dolphin in the Florida Keys." *Gulf of Mexico Science* 21 (2003): 92–97.

Licht, Thomas. "Discriminating between Hungry and Satiated Predators: The Response of Guppies (Poecilia *reticulata*) from High and Low Predation Sites." *Ethology* 82 (1989): 238–43.

Lindstrom, Kai, and R. Craig Sargent. "Food Access, Brood Size and Filial Cannibalism in the Fantail Darter, *Etheostoma flabellare*." *Behavioral Ecology and Sociobiology* 40 (1997): 107–10.

Linnaeus, Carl. *Systema naturæ per regna tria naturæ, secundum classes, ordines, genera, species, cum characteribus, differentiis, synonymis, locis.* 10th ed., rev. Vol. 1. Stockholm: Laurentius Salvius, 1758.

Longfellow, Henry W. *The Courtship of Miles Standish, and Other Poems.* Boston: Ticknor and Fields, 1859.

———. *The Song of Hiawatha.* Boston: Ticknor and Fields, 1855.

Loppnow, Grace L., Kris Vascotto, and Paul A. Venturelli. "Invasive Smallmouth Bass (*Micropterus dolomieu*): History, Impacts, and Control." *Management of Biological Invasions* 4 (2004): 191–206.

Lougheed, Vanessa L., Barb Crosbie, and Patricia Chow-Fraser. "Predictions on the Effect of Common Carp (*Cyprinus carpio*) Exclusion on Water Quality, Zooplankton, and Submergent Macrophytes in a Great Lakes Wetland." *Canadian Journal of Fisheries and Aquatic Sciences* 55 (1998): 1189–97.

Lundberg, John G., and Edie Marsh. "Evolution and Functional Anatomy of the Pectoral Fin Rays in Cyprinoid Fishes: With Emphasis on the Suckers (Family Catostomidae)." *American Midland Naturalist* 96 (1976): 332–49.

Magee, Sarah E., and Bryan D. Neff. "Temporal Variation in Decisions about Parental Care in Bluegill, *Lepomis macrochirus*." *Ethology* 112 (2006): 1000–1007.

Magurran, Anne E. "Individual Differences in Fish Behaviour." In *The Behaviour of Teleost Fishes,* edited by T. J. Pitcher, 338–65. London: Croom Helm, 1986.

Magurran, Anne E., and Sarah L. Girling. "Predator Model Recognition and Response Habituation in Shoaling Minnows." *Animal Behaviour* 34 (1986): 510–18.

Magurran, Anne E., and Anthony Higham. "Information Transfer across Fish Shoals under Predator Threat." *Ethology* 78 (1988): 153–58.

Magurran, Anne E., and T. J. Pitcher. "Provenance, Shoal Size and the Sociobiology of Predator Evasive Behaviour in Minnow Shoals." *Proceedings of the Royal Society* 229B (1987): 439–65.

Mai, J., and J. E. Kinsella. "Composition of Lipids and Proteins of Deboned Minced and Filleted White Sucker (*Catostomus commersoni*)." *Journal of Food Biochemistry* 3 (1980): 229–39.

Mallis, Rachael E., and Lynne K. Rieske. "Arboreal Spiders in Eastern Hemlock." *Environmental Entomology* 40 (2011): 1378–87.

Markle, Douglas F. "Audubon's Hoax: Ohio River Fishes Described by Rafinesque." *Archives of Natural History* 24 (1997): 439–47.

Martin, Katherine L., and P. Charles Goebel. "The Foundation Species Influence of Eastern Hemlock (*Tsuga canadensis*) on Biodiversity and Ecosystem Function on the Unglaciated Allegheny Plateau." *Forest Ecology and Management* 289 (2013): 143–52.

Matthews, William J. "Critical Current Speeds and Microhabitats of the Benthic Fishes *Percina roanoka* and *Etheostoma flabellare*." *Environmental Biology of Fishes* 12 (1985): 303–8.

Matthews, William J., Mary E. Power, and Arthur J. Stewart. "Depth Distribution of *Campostoma* Grazing Scars in an Ozark Stream." *Environmental Biology of Fishes* 17 (1986): 291–97.

Maurakis, Eugene G. "Breeding Behaviors in *Nocomis platyrhynchus* and *Nocomis raneyi* (Actinopterygii: Cyprinidae)." *Virginia Journal of Science* 49 (1998): 227–36.

Maurakis, Eugene G., and Terre D. Green. "Comparison of Spawning and Non-spawning Substrates in Nests of Species of *Exoglossum* and *Nocomis* (Actinopterygii: Cyprinidae)." *Virginia Journal of Science* 52 (2001): 25–34.

Maurakis, Eugene G., William S. Woolcott, and Mark H. Sabaj. "Reproductive-Behavioral Phylogenetics of *Nocomis* Species-Groups." *American Midland Naturalist* 126 (1991): 103–10.

Maurer, Fr. *Die epidermis und ihre abkömmlinge.* Leipzig: Wilhelm Engelmann, 1895.

Maynard Smith, John. *Evolution and the Theory of Games.* Cambridge: Cambridge University Press, 1982.

Maynard Smith, J., and G. R. Price. "The Logic of Animal Conflict." *Nature* 246 (1973): 15–18.

McGuire, William R. "Relationships of Epidermal Morphology and Breeding Behaviors in Pebble Nest-Building Minnows (Pisces: Cyprinidae)." MS thesis, University of Richmond, 1993.

McKinstry, Donald M. "Catfish Stings in the United States: Case Report and Review." *Wilderness and Environmental Medicine* 4 (1993): 293–303.

Mead, Joel K. *The National Register, vol. 4, no. 8.* Washington City: Printed and Published by the Proprietor, 1817.

Meehan, William E. *Fish, Fishing, and Fisheries of Pennsylvania.* Harrisburg, PA: E. K. Meyers, State Printer, 1893.

———. "To the Committee of Fish and Game of the Blooming Grove Park Association." In *Report of the Commissioners of Fisheries of the State of Pennsylvania for the Year 1900,* 138. Harrisburg: Wm. Stanley Ray, State Printer of Pennsylvania, 1900.

Meisner, J. Donald. "Effect of Climatic Warming on the Southern Margins of the Native Range of Brook Trout, *Salvelinus fontinalis.*" *Canadian Journal of Fisheries and Aquatic Sciences* 47, no. 6 (1990): 1065–70.

Menounos, Brian, John J. Clague, Gerald Osborn, Brian H. Luckman, Thomas R. Lakeman, and Ryan Minkus. "Western Canadian Glaciers Advance in Concert with Climate Change circa 4.2 ka." *Geophysical Research Letters* 35, no. 7 (2008). https://doi.org/10.1029/2008GL033172.

Merriam, C. Hart. "Baird the Naturalist." *Scientific Monthly* 18 (1924): 588–95.

Meryman, Richard. "A Painter of Angels Became the Father of Camouflage." *Smithsonian* 30, no. 1 (1999): 116–29.

Milanković, Milutin. "Mathematische klimalehre und astronomische theorie der klimaschwankungen." In *Handbuch der Klimatologie,* vol. 1, part A, edited by W. Köppen and R. Geiger, 1–176. Berlin: Borntraeger, 1930.

Milinski, Manfred. "Risk of Predation of Parasitized Sticklebacks (*Gasterosteus aculeatus* L.) under Competition for Food." *Behaviour* 93 (1985): 203–16.

———. "Tit for Tat in Sticklebacks and the Evolution of Cooperation." *Nature* 325 (1987): 433–35.

Milinski, Manfred, and Theo C. M. Bakker. "Female Sticklebacks Use Male Coloration in Mate Choice and Hence Avoid Parasitized Males." *Nature* 344 (1990): 330–33.

Milinski, Manfred, David Kulling, and Rolf Kettler. "Tit for Tat: Sticklebacks 'Trusting' a Cooperating Partner." *Behavioral Ecology* 1 (1990): 7–11.

Miller, Helen C. "The Behavior of the Pumpkinseed Sunfish, *Lepomis gibbosus* (Linneaus), with Notes on the Behavior of Other Species of *Lepomis* and the Pigmy Sunfish, *Elassoma evergladei*." *Behaviour* 22 (1963): 88–151.

Miller, Noam Y., and Robert Gerlai. "Shoaling in Zebrafish: What We Don't Know." *Reviews in the Neurosciences* 22 (2011): 17–25.

Miller, Rudolph J. "Behavior and Ecology of Some North American Cyprinid Fishes." *American Midland Naturalist* 72 (1964): 313–57.

———. "Reproductive Behavior of the Stoneroller Minnow, *Campostoma anomalum pullum*." *Copeia* 1962 (1962): 407–417.

Miller, Rudolph J., and Howard E. Evans. "External Morphology of the Brain and Lips in Catostomid Fishes." *Copeia* 1965 (1965): 467–87.

Miller, S. A., and T. A. Crowl. "Effects of Common Carp (*Cyprinus carpio*) on Macrophytes and Invertebrate Communities in a Shallow Lake." *Freshwater Biology* 51 (2006): 85–94.

Mills, L. Scott, Michael E. Soulé, and Daniel F. Doak. "The Keystone-Species Concept in Ecology and Conservation." *BioScience* 43 (1993): 219–24.

Milner, James W. "The Progress of Fish Culture in the United States." In *United States Commission of Fish and Fisheries Report of the Commissioner for 1872 and 1873*, 523–58. Washington, DC: US Government Printing Office, 1874.

Mishra, P. K. "Role of Smokes in Warfare." *Defence Science Journal* 44 (1994): 173–79.

Mitchill, Samuel L. "Memoir on Ichthyology: The Fishes of New York, Described and Arranged." *American Monthly Magazine and Critical Review* 2 (1818): 321–28.

———. *Report, in Part, on the Fishes of New York, 1814*. Reprinted and edited by Theodore Gill. Washington, DC: Printed for the editor, 1898.

Moore, Jonathan W. "Animal Ecosystem Engineers in Streams." *Bioscience* 56 (2006): 237–46.

Morrell, Lesley J., Graeme D. Ruxton, and Richard James. "Spatial Positioning in the Selfish Herd." *Behavioral Ecology* 22 (2010): 16–22.

Morris, D. M., and L. E. Dawson. "Storage Stability of Mechanically Deboned Sucker (Catostomidae) Flesh." *Journal of Food Science* 44 (1979): 1093–96.

Mount, D. R., C. G. Ingersoll, D. D. Gulley, J. D. Fernandez, T. W. LaPoint, and H. L. Bergman. "Effect of Long-Term Exposure to Acid, Aluminum, and Low Calcium on Adult Brook Trout (*Salvelinus fontinalis*): 1. Survival, Growth,

Fecundity, and Progeny Survival." *Canadian Journal of Fisheries and Aquatic Sciences* 45 (1988): 1623–32.

Muller, Corsin F., and Marta B. Manser. "'Nasty Neighbours' Rather than 'Dear Enemies' in a Social Carnivore." *Proceedings of the Royal Society B* 27 (2007): 959–65.

Murphy, K. E., and T. J. Pitcher. "Individual Behavioural Strategies Associated with Predator Inspection in Minnow Shoals." *Ethology* 88 (1991): 307–19.

———. "Predator Attack Motivation Influences the Inspection Behaviour of European Minnows." *Journal of Fish Biology* 50 (1997): 407–17.

Naiman, Robert J., Robert E. Bilby, Daniel E. Schindler, and James M. Helfield. "Pacific Salmon, Nutrients, and the Dynamics of Freshwater and Riparian Ecosystems." *Ecosystems* 5 (2002): 399–417.

Neff, Bryan D. "Paternity and Condition Affect Cannibalistic Behavior in Nest-Tending Bluegill Sunfish." *Behavioral Ecology and Sociobiology* 54 (2003): 377–84.

Neff, Bryan D., and Paul W. Sherman. "Nestling Recognition via Direct Cues by Parental Male Bluegill Sunfish (*Lepomis macrochirus*)." *Animal Cognition* 6 (2003): 87–92.

Nilsson, Mark C. "Market Alternatives for Carp, Alewife, and Sucker from the Michigan Great Lakes." Report Prepared for Michigan State University, 1979. https://ageconsearch.umn.edu/record/11112.

Nolde, Jack E., and William F. Giannini. "Ancient Warm Springs Deposits in Bath and Rockingham Counties, Virginia." *Virginia Minerals* 43 (1997): 9–16.

Nutter, Larry J. "Hydrogeology of Antietam Creek Basin." *Journal of Research, U.S. Geological Survey* 2 (1974): 249–52.

O'Brien, Shannon M., and Douglas E. Facey. "Habitat Use by the Eastern Sand Darter, *Ammocrypta pellucida*, in Two Lake Champlain Tributaries." *Canadian Field-Naturalist* 122 (2008): 239–46.

Obrochta, Stephen P., Hiroko Miyahara, Yusuke Yokoyama, and Thomas J. Crowley. "A Re-examination of Evidence for the North Atlantic '1500-Year Cycle' at Site 609." *Quaternary Science Reviews* 55 (2012): 23–33.

Ohio Fish Commission. *Sixth Annual Report of the Ohio Fish Commission, Made to the Governor of the State of Ohio, for the Year 1881.* Executive documents, Annual Reports for 1881, pt. 2, 1425–43. Columbus: State of Ohio, 1882.

Oliveira, Kenneth. "Life History Characteristics and Strategies of the American Eel, *Anguilla rostrata.*" *Canadian Journal of Fisheries Aquatic Sciences* 56 (1999): 795–802.

Ollerton, Jeff. "'Biological Barter': Patterns of Specialization Compared across Different Mutualisms." In *Plant-Pollinator Interactions: From Specialization to Generalization,* 411–35. Chicago: University of Chicago Press, 2006.

Page, Lawrence M., and Henry L. Bart Jr. "Egg-Mimics in Darters (Pisces: Percidae)." *Copeia* 1989 (1989): 514–17.

Page, Lawrence M., and Brooks M. Burr. *Peterson Field Guide to Freshwater Fishes of North America North of Mexico.* 2nd ed. New York: Houghton Mifflin Harcourt, 2011.

————. "The Smallest Species of Darter (Pisces: Percidae)." *American Midland Naturalist* 101 (1979): 452–53.

Page, Lawrence M., Héctor Espinosa-Pérez, Lloyd T. Findley, Carter R. Gilbert, Robert N. Lea, Nicholas E. Mandrak, Richard L. Mayden, and Joseph S. Nelson. *Common and Scientific Names of Fishes from the United States, Canada, and Mexico.* Special Publication 34. Bethesda, MD: American Fisheries Society, 2013.

Page, Lawrence M., and Carol E. Johnston. "Spawning in the Creek Chubsucker, *Erimyzon oblongus,* with a Review of Spawning Behavior in Suckers (Catostomidae)." *Environmental Biology of Fishes* 27 (1990): 265–72.

Page, Lawrene M., and David L. Swofford. "Morphological Correlates of Ecological Specialization in Darters." *Environmental Biology of Fishes* 2 (1984): 139–59.

Pakarian, Pooya, and Mohammad Taghi Yasamy. "Wagon-Wheel Illusion under Steady Illumination: Real or Illusory?" *Perception* 32 (2003): 1307–10.

Parkos, Joseph J., III, Victor J. Santucci Jr., and David H. Wahl. "Effects of Adult Common Carp (*Cyprinus carpio*) on Multiple Trophic Levels in Shallow Mesocosms." *Canadian Journal of Fisheries and Aquatic Sciences* 60 (2003): 182–92.

Penczek, John, Paul A. Boynton, and Jolene D. Splett. "Color Error in the Digital Camera Image Capture Process." *Journal of Digital Imaging* 27 (2014): 182–91.

Pennant, Thomas. *Arctic Zoology.* Vol. 1. London: Henry Hughs, 1784.

Peoples, Brandon K., Ryan A. McManamay, Donald J. Orth, and Emmanuel A. Frimpong. "Nesting Habitat Use by River Chubs in a Hydrologically Variable Appalachian Tailwater." *Ecology of Freshwater Fish* 23 (2014): 283–93.

Petty, J. Todd, Jeff L. Hansbarger, Brock M. Huntsman, and Patricia M. Mazik. "Brook Trout Movement in Response to Temperature, Flow, and Thermal Refugia within a Complex Appalachian Riverscape." *Transactions of the American Fisheries Society* 141 (2012): 1060–73.

Pflieger, William L. *Fishes of Missouri.* [Jefferson City]: Missouri Department of Conservation, 1975.

Pitcher, Tony J. "Who Dares, Wins: The Functions and Evolution of Predator Inspection Behaviour in Shoaling Fish." *Netherlands Journal of Zoology* 42 (1992): 371–91.

Pitcher, T. J., D. A. Green, and A. E. Magurran. "Dicing with Death: Predator Inspection Behaviour in Minnow Shoals." *Journal of Fish Biology* 28 (1986): 438–48.

Pitcher, Tony J., and Julia K. Parrish. "Functions of Shoaling Behaviour in Teleosts." In *The Behaviour of Teleost Fishes,* 2nd ed., edited by Tony J. Pitcher, 364–439. London: Croom Helm, 1993.

Poly, William J. *Family Percopsidae Agassiz 1850—Troutperches and Sand Rollers: Annotated Checklist of Fishes, no. 23.* San Francisco: California Academy of Sciences, 2004.

Poplar-Jeffers, Ira O., J. Todd Petty, James T. Anderson, Steven J. Kite, Michael P. Strager, and Ronald H. Fortney. "Culvert Replacement and Stream Habitat

Restoration: Implications from Brook Trout Management in an Appalachian Watershed, U.S.A." *Restoration Ecology* 17 (2009): 404–13.

Potter, Ian C. "Ecology of Larval and Metamorphosing Lampreys." *Canadian Journal of Fisheries and Aquatic Sciences* 37 (1980): 1641–57.

Potter, Ian C., Howard S. Gill, and Claude B. Renaud. "Petromyzontidae: Lampreys." In *Freshwater Fishes of North America,* edited by Melvin L. Warren Jr. and Brooks M. Burr, 105–39. Baltimore: Johns Hopkins University Press, 2014.

Power, Mary E., and William J. Matthews. "Algae-Grazing Minnows (*Campostoma anomalum*), Piscivorous Bass (*Micropterus* spp.), and the Distribution of Attached Algae in a Small Prairie-Margin Stream." *Oecologia* 60 (1983): 328–32.

Prandtl, Ludwig. "Über flüssigkeitsbewegung bei sehr kleiner reibung." In *Verhandlungen des III: Internationalen Mathematiker-Kongresses in Heidelberg,* 484–91. Leipzig: B. G. Teubner, 1905.

Preisser, Evan L., Kelly L. F. Oten, and Fred P. Hain. "Hemlock Woolly Adelgid in the Eastern United States: What Have We Learned?" *Southeastern Naturalist* 13 (2014): 1–15.

Pruett-Jones, Stephen. "Independent versus Nonindependent Mate Choice: Do Females Copy Each Other?" *American Naturalist* 140 (1992): 1000–1009.

Rafinesque, Constantine S. "Descriptions of Two New Genera of North American Fishes, *Opsanus* and *Notropis.*" *American Monthly Magazine and Critical Review* 2 (1818): 203–4.

———. *Ichthyologia Ohiensis or Natural History of the Stream Fishes Inhabiting the River Ohio and Its Natural Tributary Streams.* Lexington, KY: W. G. Hunt, 1820.

———. "On 3 N. Sp. of *Clintonia.*" *Atlantic Journal* 1 (1832): 120.

Ray, A. K., K. Ghosh, and E. Ringø. "Enzyme-Producing Bacteria Isolated from Fish Gut: A Review." *Aquaculture Nutrition* 18 (2012): 465–92.

Reed, Hugh D. "The Poison Glands of *Noturus* and *Schilbeodes.*" *American Naturalist* 41 (1907): 553–66.

Reighard, Jacob. "The Breeding Behavior of the Suckers and Minnows." *Biological Bulletin* 38, no. 1 (1920): 1–32.

———. "The Breeding Habits of the River Chub, *Nocomis micropogon* (Cope)." *Papers of the Michigan Academy of Science, Arts and Letters* 28 (1943): 393–423.

———. "The Function of the Pearl Organs of Cyprinidae." *Science* 17 (1903): 531.

———. "Methods of Studying the Habits of Fishes, with an Account of the Breeding Habits of the Horned Dace." *Bulletin of the U.S. Bureau of Fisheries* 28 (1910): 1111–36.

———. "The Pearl Organs of American Minnows in Their Relation to the Factors of Descent." *Science,* n.s., 31 (1910): 472.

Reynolds, Osborne. "An Experimental Investigation of the Circumstances Which Determine Whether the Motion of Water Shall Be Direct or Sinuous, and of the Law of Resistance in Parallel Channels." *Philosophical Transactions of the Royal Society A* 174 (1883): 935–82.

———. "Experiments Showing Dilatancy, a Property of Granular Material, Possibly Connected with Gravitation." In *Notices of the Proceedings at the Meeting of the Members of the Royal Institution of Great Britain with Abstracts of the Discourses Delivered at the Evening Meetings,* vol. 9, *1884–1886,* 354–63. London: William Clowes and Sons, 1887.

———. *On an Inversion of Ideas as to the Structure of the Universe.* Rede Lecture, June 10, 1902. Cambridge: Cambridge University Press, 1903.

———. "On the Dilatancy of Media Composed of Rigid Particles in Contact: With Experimental Illustrations." *London, Edinburgh, and Dublin Philosophical Magazine and Journal of Science* 20 (1885): 469–81.

———. *Papers on Mechanical and Physical Subjects.* Vol. 3, *The Sub-mechanics of the Universe.* Cambridge: Cambridge University Press, 1903.

Rimmer, D. W., and W. J. Wiebe. "Fermentative Microbial Digestion in Herbivorous Fishes." *Journal of Fish Biology* 31 (1987): 229–36.

Roberts, Tyson R. "Unculi (Horny Projections Arising from Single Cells), an Adaptive Feature of the Epidermis of Ostariophysan Fishes." *Zoologica Scripta* 11 (1982): 55–76.

Robins, C. Richard. "*Cottus kanawhae,* a New Cottid Fish from the New River System of Virginia and West Virginia." *Zootaxa* 987 (2005): 1–6.

Robins, C. Richard, Reeve M. Bailey, Carl E. Bond, James R. Booker, Ernest A. Lachner, Robert N. Lea, and W. B. Scott. *Common and Scientific Names of Fishes from the United States and Canada,* 5th ed. Bethesda, MD: American Fisheries Society, 1991.

Robison, Henry W., and Thomas M. Buchanan. *Fishes of Arkansas.* Fayetteville: University of Arkansas Press, 1988.

Rogick, Mary D. "Studies on the Comparative Histology of the Digestive Tube of Certain Teleost Fishes, II. A Minnow (*Campostoma anomalum*)." *Journal of Morphology and Physiology* 52 (1931): 1–25.

Rohr, Jason R., Carolyn G. Mahan, and Ke Chung Kim. "Response of Arthropod Biodiversity to Foundation Species Declines: The Case of the Eastern Hemlock." *Forest Ecology and Management* 258 (2009): 1503–10.

Roosevelt, Theodore. "Revealing and Concealing Coloration in Birds and Mammals." *Bulletin of the American Museum of Natural History* 30 (1911): 120–221.

Ross, Michael R. "The Breeding Behavior and Hybrid Potential of the Northern Creek Chub *Semotilus atromaculatus atromaculatus* (Mitchill)." PhD diss., Ohio State University, 1975.

———. "Function of Creek Chub (*Semotilus atromaculatus*) Nest-Building." *Ohio Journal of Science* 77 (1977): 36–37.

Ross, R. M., R. M. Bennett, C. D. Snyder, J. A. Young, D. R. Smith, and D. P. Lemarié. "Influence of Eastern Hemlock (*Tsuga canadensis* L.) on Fish Community Structure and Function in Headwater Streams of the Delaware River Basin." *Ecology of Freshwater Fish* 12 (2003): 60–65.

Ross, Michael R., and Roger J. Reed. "The Reproductive Behavior of the Fallfish *Semotilus corporalis.*" *Copeia* 1978 (1978): 215–21.

Rossi-Santos, Marcos R., and Leonardo L. Wedekin. "Evidence of Bottom Contact Behavior by Estuarine Dolphins (*Sotalia guianensis*) on the Eastern Coast of Brazil." *Aquatic Mammals* 32 (2006): 140–44.

Rott, N. "Note on the History of the Reynolds Number." *Annual Review of Fluid Mechanics* 22 (1990): 1–11.

Rowher, Sievert. "Parent Cannibalism of Offspring and Egg Raiding as a Courtship Strategy." *American Naturalist* 112 (1978): 429–40.

Ruble, Crystal L. "Captive Propagation, Reproductive Biology, and Early Life History of *Crystallaria cincotta* (Diamond Darter), *Etheostoma wapiti* (Boulder Darter), *E. vulneratum* (Wounded Darter), and *E. maculatum* (Spotted Darter)." MS thesis, West Virginia University, 2013.

Ruble, Crystal L., Patrick L. Rakes, John R. Shute, and Stuart A. Welsh. "Captive Propagation, Reproductive Biology, and Early Life History of the Diamond Darter (*Crystallaria cincotta*)." *American Midland Naturalist* 172 (2014): 107–18.

Ruxton, Graeme D. "The Possible Fitness Benefits of Striped Coat Coloration for Zebra." *Mammal Review* 32 (2002): 237–44.

Ruxton, Graeme D., Michael P. Speed, and David J. Kelly. "What, If Anything, Is the Adaptive Function of Countershading?" *Animal Behaviour* 68 (2004): 445–51.

Saad, David A., and Daniel J. Hippe. *Large Springs in the Valley and Ridge Physiographic Province of Pennsylvania.* US Geological Survey Open-File Report 90-164. Harrisburg, PA: US Geological Survey, 1990.

Sabaj, Mark H. "Spawning Clasps and Gamete Deposition in Pebble Nest-Building Minnows (Pisces: Cyprinidae)." MS thesis, University of Richmond, 1992.

Sabaj, Mark H., Eugene G. Maurakis, and William S. Woolcott. "Spawning Behaviors in the Bluehead Chub, *Nocomis leptocephalus*, River Chub, *N. micropogon*, and Central Stoneroller, *Campostoma anomalum*." *American Midland Naturalist* 144 (2000): 187–201.

Sabino, José, Luciana P. Andrade, Ivan Sazima, Fabricio B. Teresa, Sergio R. Floeter, Cristina Sazima, and Roberta M. Bonaldo. "Following Fish Feeding Associations in Marine and Freshwater Habitats." *Marine and Freshwater Research* 68 (2017): 381–87.

Sargent, R. Craig. "Paternal Care and Egg Survival Both Increase with Clutch Size in the Fathead Minnow, *Pimephales promelas*." *Behavioral Ecology and Sociobiology* 23 (1988): 33–37.

Scarnecchia, Dennis L. "A Reappraisal of Gars and Bowfins in Fishery Management." *Fisheries* 17 (1992): 6–12.

Schaefer, H. Martin, and Nina Stobbe. "Disruptive Coloration Provides Camouflage Independent of Background Matching." *Proceedings of the Royal Society B* 273 (2006): 2427–32.

Schmidly, David J. "What It Means to Be a Naturalist and the Future of Natural History at American Universities." *Journal of Mammalogy* 86 (2005): 449–56.

Schmitt, Francois G. "Turbulence from 1870 to 1920: The Birth of a Noun and of a Concept." *Comptes Rendus Mécanique* 345 (2017): 620–26.

Schriefer, Julie E., and Melina E. Hale. "Strikes and Startles of Northern Pike (*Esox lucius*): A Comparison of Muscle Activity and Kinematics between S-Start Behaviors." *Journal of Experimental Biology* 207 (2004): 535–44.

Schultz, Robert A., William A. Hobba Jr., and Mark D. Kozar. *Geohydrology, Ground-Water Availability, and Ground-Water Quality of Berkeley County, West Virginia, with Emphasis on the Carbonate-Rock Area.* Water-Resources Investigations Report 93-4073. Charleston, WV: US Geological Survey, 1995.

Scott, W. B., and E. J. Crossman. *Freshwater Fishes of Canada.* Bulletin 184. Ottawa: Fisheries Research Board of Canada, 1973.

Segall, Harold N., William H. Helfand, and Lloyd. G. Stevenson. "Historic Marker for Charles Conrad Abbott Farm." *Journal of the History of Medicine and Allied Sciences* 22 (1967): 418–19.

Seuss, Dr. *Oh Say Can You Say?* New York: Random House, 1979.

Seversmith, Herbert F. "Distribution, Morphology, and Life History of *Lampetra aepyptera*, a Brook Lamprey, in Maryland." *Copeia* 1953 (1953): 225–32.

Shaler, Nathaniel S. *Nature and Man in America.* New York: Charles Scribner's Sons, 1897.

Shaler, Nathaniel S., and Sophia Penn Page Shaler. *The Autobiography of Nathaniel Southgate Shaler with a Supplementary Memoir by His Wife.* Boston: Houghton Mifflin, 1909.

Shuman, Bryan N., Patrick J. Bartlein, and Thompson Webb III. "The Magnitudes of Millennial- and Orbital-Scale Climatic Change in Eastern North America during the Late Quaternary." *Quaternary Science Reviews* 24 (2005): 2194–206.

Shuman, Bryan N., and Jeremiah Marsicek. "The Structure of Holocene Climate Change in Mid-latitude North America." *Quaternary Science Reviews* 141 (2016): 38–51.

Shuman, Bryan N., Paige Newby, and Jeffrey P. Donnelly. "Abrupt Climate Change as an Important Agent of Ecological Change in the Northeast U.S. throughout the Past 15,000 Years." *Quaternary Science Reviews* 28 (2009): 1693–709.

Siderhurst, Leigh A., Heather P. Griscom, Mark Hudy, and Zachary J. Bortolot. "Changes in Light Levels and Stream Temperatures with Loss of Eastern Hemlock (*Tsuga canadensis*) at a Southern Appalachian Stream: Implications for Brook Trout." *Forest Ecology and Management* 260 (2010): 1677–88.

Sih, Andrew. "Optimal Behavior: Can Foragers Balance Two Conflicting Demands?" *Science* 210 (1980): 1041–43.

Simon, Thomas P. "Startle Response and Causes of Burying Behavior in Captive Eastern Sand Darters, *Ammocrypta pellucida* (Putnam)." *Proceedings of the Indiana Academy of Science* 100 (1991): 155–60.

Sivan, Gisha. "Fish Venom: Pharmacological Features and Biological Significance." *Fish and Fisheries* 10 (2009): 159–72.

Smiley, Charles W. "Answers Relative to 118 Questions Relative to German Carp." *Bulletin of the United States Fish Commission* 3 (1883): 241–48.

———. "Brief Notes upon Fish and the Fisheries." *Bulletin of the United States Fish Commission* 4 (1884): 305–20.

———. "Notes on the Edible Qualities of German Carp and Hints about Cooking Them." *Bulletin of the United States Fish Commission* 3 (1883): 305–32.

Smith, Dustin M. "Habitat Selection and Predation Risk in Larval Lampreys." MS thesis, West Virginia University, 2009.

Smith, Dustin M., Stuart A. Welsh, and Philip J. Turk. "Available Benthic Habitat Type May Influence Predation Risk in Larval Lampreys." *Ecology of Freshwater Fish* 21 (2012): 160–63.

———. "Selection and Preference of Benthic Habitat by Small and Large Ammocoetes of the Least Brook Lamprey (*Lampetra aepyptera*)." *Environmental Biology of Fishes* 91 (2011): 421–28.

Smith, Osgood R. "The Breeding Habits of the Stone Roller Minnow (*Campostoma anomalum* Rafinesque)." *Transactions of the American Fisheries Society* 65 (1935): 148–51.

Smith, William L., and Ward C. Wheeler. "Venom Evolution Widespread in Fishes: A Phylogenetic Road Map for the Bioprospecting of Piscine Venoms." *Journal of Heredity* 97 (2006): 206–17.

Snyder, Craig D., John A. Young, David P. Lemarié, and David R. Smith. "Influence of Eastern Hemlock (*Tsuga canadensis*) Forests on Aquatic Invertebrate Assemblages in Headwater Streams." *Canadian Journal of Fisheries and Aquatic Sciences* 59 (2002): 262–75.

Spreitzer, August E. "Life History, External Morphology, Osteology of the Eastern Sand Darter *Ammocrypta pellucida* (Putnam, 1863), an Endangered Ohio Species (Pisces: Percidae)." MS thesis, Ohio State University, 1979.

Stevens, Martin. "The Role of Eyespots as Anti-predator Mechanisms, Principally Demonstrated in the Lepidoptera." *Biological Reviews* 80 (2005): 573–88.

Stevens, Martin, Innes C. Cuthill, C. Alejandro Parraga, and Tom Troscianko. "The Effectiveness of Disruptive Coloration as a Concealment Strategy." *Progress in Brain Research* 155 (2006): 49–64.

Stevens, Martin, Chloe J. Hardman, and Claire L. Stubbins. "Conspicuousness, Not Eye Mimicry, Makes 'Eyespots' Effective Antipredator Signals." *Behavioral Ecology* 19 (2008): 525–31.

Stevens, Martin, Elinor Hopkins, William Hinde, Amabel Adcock, Yvonne Connolly, Tom Troscianko, and Innes C. Cuthill. "Field Experiments on the Effectiveness of 'Eyespots' as Predator Deterrents." *Animal Behaviour* 74 (2007): 1215–27.

Stevens, Martin, and Graeme D. Ruxton. "Do Animal Eyespots Really Mimic Eyes?" *Current Zoology* 60 (2014): 26–36.

Stevens, Martin, W. Tom L. Searle, Jenny E. Seymour, Kate L. Marshall, and Graeme D. Ruxton. "Motion Dazzle and Camouflage as Distinct Anti-predator Defenses." *BMC Biology* 9 (2011): 81.

Stewart, Norman H. "Pearl Organs of the Cyprinidae." *Proceedings of the Pennsylvania Academy of Science* 27 (1953): 221–24.

Strandberg, Carl H., and Ray Tomlinson. "Photoarchaeological Analysis of Potomac River Fish Traps." *American Antiquity* 34 (1969): 312–19.

Stranko, Scott A., Robert H. Hilderbrand, Raymond P. Morgan II, Mark W. Staley, Andrew J. Becker, Ann Roseberry-Lincoln, Elgon S. Perry, and Paul T. Jacobson. "Brook Trout Declines with Land Cover and Temperature Changes in Maryland." *North American Journal of Fisheries Management* 28 (2008): 1223–32.

Studinski, Jered M., Kyle J. Hartman, Jonathan M. Niles, and Patrick Keyser. "The Effects of Riparian Forest Disturbance on Stream Temperature, Sedimentation, and Morphology." *Hydrobiologia* 686 (2012): 107–17.

Surface, H. A. "Removal of Lampreys from the Interior Waters of New York." In *Fourth Annual Report, Commission of Fisheries, Game and Forests of the State of New York,* 191–245. Albany, NY: Wynkoop Hallenbeck Crawford, 1899.

Sutton, Mark Q. "The 'Fishing Link': Salmonids and the Initial Peopling of the Americas." *PaleoAmerica* 3 (2017): 231–59.

Sweka, John A., and Kyle J. Hartman. "Contribution of Terrestrial Invertebrates to Yearly Brook Trout Prey Consumption and Growth." *Transactions of the American Fisheries Society* 137 (2008): 224–35.

———. "Effects of Large Woody Debris Addition on Stream Habitat and Brook Trout Populations in Appalachian Streams." *Hydrobiologia* 559 (2006): 363–78.

Switzer, John F., Stuart A. Welsh, and Tim L. King. "Microsatellite DNA Primers for the Candy Darter, *Etheostoma osburni* and Variegate Darter, *Etheostoma variatum,* and Cross-Species Amplification in Other Darters (Percidae)." *Molecular Ecology Resources* 8 (2008): 335–38.

———. "A Molecular Genetic Investigation of Hybridization between *Etheostoma osburni* and *Etheostoma variatum* in the New River Drainage, West Virginia." Report Submitted to the West Virginia Division of Natural Resources, Elkins, WV, 2007.

Tarr, Ralph S. "The United States Fish Commission." *Nature* 31 (1884): 128–30.

Taylor, William R. *A Revision of the Catfish Genus Noturus Rafinesque, with an Analysis of Higher Groups in the Ictaluridae.* Bulletin of the United States National Museum, no. 282. Washington, DC: Smithsonian Institution Press, 1969.

Templeton, Alan R., Kerry Shaw, Eric Routman, and Scott K. Davis. "The Genetic Consequences of Habitat Fragmentation." *Annals of the Missouri Botanical Garden* 77 (1990): 13–27.

Teresa, Fabricio B., and Fernando R. Carvalho. "Feeding Association between Benthic and Nektonic Neotropical Stream Fishes." *Neotropical Ichthyology* 6 (2008): 109–11.

Teresa, Fabricio B., Cristina Sazima, Ivan Sazima, and Sergio R. Floeter. "Predictive Factors of Species Composition of Follower Fishes in Nuclear-Follower Feeding Associations: A Snapshot Study." *Neotropical Ichthyology* 12 (2014): 913–19.

Thayer, Abbott H. "The Law Which Underlies Protective Coloration." *Auk* 13 (1896): 124–29.

———. "Protective Coloration in Its Relation to Mimicry, Common Warming Colours, and Sexual Selection." *Transactions of the Entomological Society of London* 51 (1903): 553–69.

Thayer, Gerald H. *Concealing-Coloration in the Animal Kingdom: An Exposition of the Laws of Disguise through Color and Pattern; Being a Summary of Abbott H. Thayer's Discoveries.* New York: Macmillan, 1909.

Thompson, Edward. "Edible Qualities of Carp." *Bulletin of the United States Fish Commission* 4 (1884): 176.

Thompson, Patricia A., Stuart A. Welsh, Austin A. Rizzo, and Dustin M. Smith. "Effect of Substrate Size on Sympatric Sand Darter Benthic Habitat Preferences." *Journal of Freshwater Ecology* 32 (2017): 455–65.

Thompson, Patricia A., Stuart A. Welsh, Michael P. Strager, and Austin A. Rizzo. "A Multiscale Investigation of Habitat Use and Within-River Distribution of Sympatric Sand Darter Species." *Journal of Geospatial Applications in Natural Resources* 2 (2018). https://scholarworks.sfasu.edu/j_of_geospatial _applications_in_natural_resources/vol2/iss1/1.

Thompson, Peter. "Does My Butt Look Big in This? Horizontal Stripes, Perceived Body Size and the Oppel-Kundt Illusion." *Journal of Vision* 8 (2008): 822.

Thompson, Peter, and Kyriaki Mikellidou. "Applying the Helmholtz Illusion to Fashion: Horizontal Stripes Won't Make You Look Fatter." *i-Perception* 2 (2011): 69–76.

———. "The 3-D Helmholtz Square Illusion: More Reasons to Wear Horizontal Stripes." *Journal of Vision* 9 (2009): 50.

Thompson, Zadock. *History of Vermont, Natural, Civil and Statistical, in Three Parts, with an Appendix.* Burlington, VT: Stacy and Jameson, 1853.

———. "Species Description of *Percopsis pellucida*." Presented to five members of the Boston Society of Natural History on July 18, 1849, by D. H. Storer, on behalf of Zadock Thompson. *Proceedings of the Boston Society of Natural History* 3 (1848–51): 163–65.

Threinen, C. W., and William T. Helm. "Experiments and Observations Designed to Show Carp Destruction of Aquatic Vegetation." *Journal of Wildlife Management* 18 (1954): 247–51.

Tiegs, Scott D., Peter S. Levi, Janine Ruegg, Dominic T. Chaloner, Jennifer L. Tank, and Gary A. Lamberti. "Ecological Effects of Live Salmon Exceed Those of Carcasses during an Annual Spawning Migration." *Ecosystems* 14 (2011): 598–614.

Timm, Anne, Eric Hallerman, C. Andrew Dolloff, Mark Hudy, and Randall Kolka. "Identification of a Barrier Height Threshold Where Brook Trout Population Genetic Diversity, Differentiation, and Relatedness Are Affected." *Environmental Biology of Fishes* 99 (2016): 195–208.

Tinbergen, Nikolaas. "The Curious Behavior of the Stickleback." *Scientific American* 187 (1952): 22–27.

———. "Ethology and Stress Diseases." *Science,* n.s., 185 (1974): 20–27.

Tobler, Michael, and Ingo Schlupp. "Influence of Black Spot Disease on Shoaling Behaviour in Female Western Mosquitofish, *Gambusia affinis* (Poeciliidae, Teleostei)." *Environmental Biology of Fishes* 81 (2008): 29–34.

Toker, Franklin. *Fallingwater Rising: Frank Lloyd Wright, E. J. Kaufmann, and America's Most Extraordinary House.* New York: Alfred A. Knopf, 2003.

Torres, Leigh G., and Andrew J. Read. "Where to Catch a Fish? The Influence of Foraging Tactics on the Ecology of Bottlenose Dolphins (*Tursiops truncatus*) in Florida Bay, Florida." *Marine Mammal Science* 25 (2009): 797–815.

Trautman, Milton B. *The Fishes of Ohio.* Columbus: Ohio State University Press, 1981.

Turcotte, Richard M. "Temporal and Spatial Distribution of Imidacloprid and the Arthropod Fauna Associated with Eastern Hemlock, *Tsuga canadensis* (L.) Carr." PhD diss., West Virginia University, 2016.

Turner, George F., and Tony J. Pitcher. "Attack Abatement: A Model for Group Protection by Combined Avoidance and Dilution." *American Naturalist* 128 (1986): 228–40.

Tyler, Jack D. "Food Habits, Sex Ratios, and Size of Longnose Gar in Southwestern Oklahoma." *Proceedings of the Oklahoma Academy of Science* 74 (1994): 41–42.

Uhler, P. R., and Otto Lugger. "List of Fishes of Maryland." In *Report of the Commissioners of Fisheries of Maryland to the General Assembly, January 1st 1876,* 67–186. Annapolis, MD: John F. Wiley, State Printer, 1876.

US Fish and Wildlife Service. "Endangered and Threatened Wildlife and Plants: Designation of Critical Habitat for the Candy Darter." *Federal Register* 83 (2018): 59232–68.

———. "Endangered and Threatened Wildlife and Plants: Endangered Species Status for Diamond Darter; Final Rule." *Federal Register* 78 (2013): 45074–95.

———. "Endangered and Threatened Wildlife and Plants: Endangered Species Status for the Candy Darter." *Federal Register* 83 (2018): 58747–54.

———. *Species Status Assessment (SSA) Report for the Tippecanoe Darter (Etheostoma tippecanoe), Version 1.0.* Hadley, MA: US Fish and Wildlife Service, 2018.

Utz, Ryan M., Brett C. Ratcliffe, Brett T. Moore, and Kyle J. Hartman. "Disproportionate Relative Importance of a Terrestrial Beetle Family (Coleoptera: Scarabaeidae) as a Prey Source for Central Appalachian Brook Trout." *Transactions of the American Fisheries Society* 136 (2007): 177–84.

Vander Zanden, M. Jake, Julian D. Olden, James H. Thorne, and Nicholas E. Mandrak. "Predicting Occurrences and Impacts of Smallmouth Bass Introductions in North Temperate Lakes." *Ecological Applications* 14 (2004): 132–48.

Van Duzer, Evelyn M. "Observations on the Breeding Habits of the Cut-Lips Minnow, *Exoglossum maxillingua.*" *Copeia* 1939 (1939): 65–75.

Van Meter, Harry. "Gigging the Long-Nosed Gar." *West Virginia Conservation* (May 1956): 6–7.

Van Snik Gray, E., J. M. Boltz, K. A. Kellogg, and J. R. Stauffer Jr. "Food Resource Partitioning by Nine Sympatric Darter Species." *Transactions of the American Fisheries Society* 126 (1997): 822–40.

Vesper, Dorothy J., Rachel V. Grand, Kristen Ward, and Joseph J. Donovan. "Geochemistry of a Spring-Dense Karst Watershed Located in a Complex Structural Setting, Appalachian Great Valley, West Virginia, USA." *Environmental Geology* 58 (2009): 667–78.

Viau, A. E., K. Gajewski, M. C. Sawada, and P. Fines. "Millennial-Scale Temperature Variations in North America during the Holocene." *Journal of Geophysical Research* 111 (2006): https://doi.org/10.1029/2005JD006031.

Virginia Fish Commission. *Annual Reports of the Fish Commissioners of the State of Virginia for the Years 1875–6 and 1876–7.* Richmond, VA: R. F. Walker, Superintendent Public Printing, 1877.

Von Helversen, Bettina, Lael J. Schooler, and Uwe Czienskowski. "Are Stripes Beneficial? Dazzle Camouflage Influences Perceived Speed and Hit Rates." *PLoS ONE* 8, no. 4 (2013): e61173. https://doi.org/10.1371/journal.pone.0061173.

Wainwright, Dylan K., and George V. Lauder. "Mucus Matters: The Slippery and Complex Surfaces of Fish." In *Functional Surfaces in Biology III,* edited by Stanislav N. Gorb and Elena V. Gorb, 223–46. Cham, Switzerland: Springer, 2017.

———. "Three-Dimensional Analysis of Scale Morphology in Bluegill Sunfish, *Lepomis macrochirus.*" *Zoology* 119 (2016): 182–95.

Walbaum, Johann J. *Petri Artedi sueci genera piscium: In quibus systema totum ichthyologiæ proponitur cum classibus, ordinibus, generum characteribus, specierum differentiis, observationibus plurimis; Redactis speciebus 242 ad genera 52; Ichthyologiæ pars III.* Greifswald, Germany: Impensis Ant. Ferdin. Röse, 1792.

Walker, Renee B., and Boyce N. Driskell. *Foragers of the Terminal Pleistocene in North America.* Lincoln: University of Nebraska Press, 2007.

Walker, Renee B., Kandace R. Detwiler, Scott C. Meeks, and Boyce N. Driskell. "Berries, Bones, and Blades: Reconstructing Late Paleoindian Subsistence Economy at Dust Cave, Alabama." *Midcontinental Journal of Archaeology* 26 (2001): 169–97.

Walton, Izaak. *The Compleat Angler or the Contemplative Man's Recreation.* London: Printed by T. Maxey for Rich. Marriot, S. Dunstans Church-Yard Fleetstreet, 1653.

———. *The Compleat Angler or the Contemplative Man's Recreation.* 2nd ed. London: Printed by T. M. for Rich. Marriot, Church-Yard Fleetstreet, 1655.

Warren, Melvin L., Jr., Brooks M. Burr, and Bernard R. Kuhajda. "Aspects of the Reproductive Biology of *Etheostoma tippecanoe* with Comments on Egg-Burying Behavior." *American Midland Naturalist* 116 (1986): 215–18.

Waters, Thomas F. "Replacement of Brook Trout by Brown Trout over 15 Years in a Minnesota Stream—Production and Abundance." *Transactions of the American Fisheries Society* 112 (1983): 137–46.

Webb, J. R., B. J. Cosby, J. N. Galloway, and G. N. Hornberger. "Acidification of Native Brook Trout Streams in Virginia." *Water Resources Research* 25 (1989): 1367–77.

Webb, Paul W. *Hydrodynamics and Energetics of Fish Propulsion.* Bulletin 190 of the Fisheries Research Board of Canada. Ottawa: Department of the Environment Fisheries and Marine Service, 1975.

Webb, Paul W., Cynthia L. Gerstner, and Scott T. Minton. "Station-Holding by the Mottled Sculpin, *Cottus bairdi* (Teleostei: Cottidae), and Other Fishes." *Copeia* 1996 (1996): 488–93.

Weed, Alfred C. *The Alligator Gar.* Chicago: Field Museum of Natural History, 1923.

Weisberg, Robert W. "Frank Lloyd Wright's Fallingwater: A Case Study in Inside-the-Box Creativity." *Creativity Research Journal* 23 (2011): 296–312.

Wellman, David I. "Post-flood Recovery and Distributions of Fishes in the New River Gorge National River and the Gauley River National Recreation Area." MS thesis, West Virginia University, 2004.

Welsh, Stuart A. "The Effects of Long-Term Insularization on Morphological Variation in the Checkered Sculpin." MS thesis, Frostburg State University, 1996.

Welsh, Stuart A., and Joni L. Aldinger. "A Semi-automated Method for Monitoring Dam Passage of Upstream Migrant Yellow-Phase American Eels." *North American Journal of Fisheries Management* 34 (2014): 702–9.

Welsh, Stuart A., Joni L. Aldinger, Melissa A. Braham, and Jennifer L. Zimmerman. "Synergistic and Singular Effects of River Discharge and Lunar Illumination on Dam Passage of Upstream Migrant Yellow-Phase American Eels." *ICES Journal of Marine Science* 73 (2016): 33–42.

Welsh, Stuart A., and Heather Liller. "Environmental Correlates of Upstream Migration of Yellow-Phase American Eels in the Potomac River Drainage." *Transactions of the American Fisheries Society* 142 (2013): 483–91.

Welsh, Stuart A., and Sue A. Perry. "Habitat Partitioning in a Community of Darters in the Elk River, West Virginia." *Environmental Biology of Fishes* 51 (1998): 411–19.

Welsh, Stuart A., Dustin M. Smith, and Nate D. Taylor. "Microhabitat Use of the Diamond Darter." *Ecology of Freshwater Fish* 22 (2013): 587–95.

Welsh, Stuart A., and Robert M. Wood. "*Crystallaria cincotta,* a New Species of Darter (Teleostei: Percidae) from the Elk River of the Ohio River Drainage, West Virginia." *Zootaxa* 1680 (2008): 62–68.

Weltman-Fahs, Maya, and Jason M. Taylor. "Hydraulic Fracturing and Brook Trout Habitat in the Marcellus Shale Region: Potential Impacts and Research Needs." *Fisheries* 38 (2013): 4–15.

Wenner, Charles A., and John A. Musick. "Fecundity and Gonad Observation of the American Eel, *Anguilla rostrata,* Migrating from Chesapeake Bay, Virginia." *Journal of the Fisheries Research Board of Canada* 31 (1974): 1387–91.

West Virginia Conservation Commission. *Report of the West Virginia Conservation Commission, 1908.* Charleston, WV: Tribune Printing Company, 1909.

Wheeler, A. "An Appraisal of the Zoology of C. S. Rafinesque." *Bulletin of Zoological Nomenclature* 45 (1988): 6–12.

Wiley, Derek, Raymond P. Morgan II, and Robert H. Hilderbrand. "Relations between Physical Habitat and American Eel Abundance in Five River Basins in Maryland." *Transactions of the American Fisheries Society* 133 (2004): 515–26.

Wiley, Martin L., and Bruce B. Collette. "Breeding Tubercles and Contact Organs in Fishes: Their Occurrence, Structure, and Significance." *Bulletin of the American Museum of Natural History* 143 (1970): 145–216.

Willacker, James J., Jr., William V. Sobczak, and Elizabeth A. Colburn. "Stream Macroinvertebrate Communities in Paired Hemlock and Deciduous Watersheds." *Northeastern Naturalist* 16 (2009): 101–12.

Williams, John W., Bryan N. Shuman, Thompson Webb III, Patrick J. Bartlein, and Phillip L. Leduc. "Late-Quaternary Vegetation Dynamics in North America: Scaling from Taxa to Biomes." *Ecological Monographs* 74 (2004): 309–34.

Williamson, William C. "On the Microscopic Structure of the Scales and Dermal Teeth of Some Ganoid and Placoid Fish." *Philosophical Transactions of the Royal Society of London* 139 (1849): 435–75.

Willson, J. M., Jr. "Florida Fur-Farming." *Bulletin of the United States Fish Commission* 17 (1897): 369–71.

Wilson, Edward O. *Sociobiology: The New Synthesis.* Cambridge, MA: Harvard University Press, 1975.

Winn, Howard E. "Comparative Reproductive Behavior and Ecology of Fourteen Species of Darters (Pisces-Percidae)." *Ecological Monographs* 28 (1958): 155–91.

Woldt, A. "Die besetzung des Stettiner Haffs mit karpfen." In *Circulare des Deutschen Fischerei-Vereins im Jahre 1881,* 232–37. Berlin: W. Moeser Hofbuchdruckerei, 1882.

Woodford, Darragh J., N. Dean Impson, Jenny A. Day, and I. Roger Bills. "The Predatory Impact of Invasive Alien Smallmouth Bass, *Micropterus dolomieu* (Teleostei: Centrarchidae), on Indigenous Fishes in a Cape Floristic Region Mountain Stream." *African Journal of Aquatic Science* 30, no. 2 (2005): 167–73.

Woodman, Neal. "Pranked by Audubon: Constantine S. Rafinesque's Description of John James Audubon's Imaginary Kentucky Mammals." *Archives of Natural History* 43 (2016): 95–108.

Worm, Ole. *Museum wormianum: Seu historia rerum rariorum; Tam naturalium, quam artificialium, tam domesticarum, quam exoticarum, quæ Hafniæ Danorum in ædibus authoris fervantur.* Leiden: Elsevir, 1655.

Wright, Jeremy J. "Diversity, Phylogenetic Distribution, and Origins of Venomous Catfishes." *BMC Evolutionary Biology* 9 (2009): 282.

———. "Evolutionary History of Venom Glands in the Siluriformes." In *Evolution of Venomous Animals and Their Toxins,* edited by P. Gopalakrishnakone and Anita Malhotra, 1–19. Dordrecht: Springer Netherlands, 2015.

Wright, Justin P., Clive G. Jones, and Alexander S. Flecker. "An Ecosystem Engineer, the Beaver, Increases Species Richness at the Landscape Scale." *Oecologia* 132 (2002): 96–101.

Wuerger, Sophie, Robert Shapley, and Nava Rubin. "'On the Visually Perceived Direction of Motion' by Hans Wallach: 60 Years Later." *Perception* 25 (1996): 1317–67.

Yager, Richard M., L. Neil Plummer, Leon J. Kaufmann, Daniel H. Doctor, David L. Nelms, and Peter Schlosser. "Comparison of Age Distributions Estimated from Environmental Tracers by Using Binary-Dilution and Numerical Models of Fractured and Folded Karst: Shenandoah Valley of Virginia and West Virginia, USA." *Hydrogeology Journal* 21 (2013): 1193–217.

Yorks, Thad E., Jennifer C. Jenkins, Donald J. Leopold, Dudley J. Raynal, and Donald A. Orwig. "Influences of Eastern Hemlock Mortality on Nutrient Cycling." In *Proceedings: Sustainable Management of Hemlock Ecosystems in Eastern North America,* edited by Katherine A. McManus, Kathleen S. Shields, and Dennis R. Souto, 126–33. Durham, NH: US Department of Agriculture, 2000.

INDEX

307